BIOLOGICALLY ACTIVE MOLECULES

Editor:
Prof. Dr. Urs Peter Schlunegger
Department of Organic Chemistry
University of Bern
Freiestrasse 3
CH-3012 Bern

BIOLOGICALLY ACTIVE MOLECULES

Identification, Characterization and Synthesis

Proceedings of a Seminar
on Chemistry of Biologically Active Compounds
and Modern Analytical Methods
Interlaken, September 5–7, 1988

Sponsored and Organized by
The Association of Swiss Chemists

Springer-Verlag
Berlin Heidelberg New York
London Paris Tokyo

Prof. Dr. Urs Peter Schlunegger
Department of Organic Chemistry
University of Bern
Freiestrasse 3
CH-3012 Bern

ISBN-13: 978-3-642-74584-3 e-ISBN-13: 978-3-642-74582-9
DOI: 10.1007/978-3-642-74582-9

Library of Congress Cataloging-in-Publication Data
Seminar on Chemistry of Biologically Active Compounds and Modern Analytical Methods (1988 : Interlaken, Switzerland) Biologically active molecules : identification, characterization, and synthesis : proceedings of a Seminar on Chemistry on Biologically Active Compounds and Modern Analytical Methods, Interlaken, September 5–7, 1988 / sponsored and organized by the Association of Swiss Chemists ; editor, Urs Peter Schlunegger. p. cm.
ISBN-13: 978-3-642-74584-3(U.S. : alk. paper)
1. Drugs–Analysis–Congresses. 2. Drugs–Synthesis–Congresses. I. Schlunegger, Urs P. II. Schweizerischer Chemiker-Verband. III. Title. RS189.S398 1988 547–dc 19 89-6159

Softcover reprint of the hardcover 1st edition 1989

2151/3140-543210 – Printed on acid-free paper

PREFACE

Over the past few years there has been a remarkable and rapid development of modern analytical methods, and the fields of nuclear magnetic resonance and mass spectrometry have been no exception. In addition to being able to do "more and faster", new innovative techniques have also arisen to contribute to a growing understanding of the relationship between chemical structure and biological activity.

In order to explore a few of the more interesting points of those developments and applications, a seminar "From Biological Activity to Structure" was organized from September 5-7, 1988 at Interlaken. The four invited speakers, Richard M. Caprioli, Howard R. Morris, Wolfgang Steglich and Dudley H. Williams were generous enough to attend and discuss many facets of their research, especially methodological and technical developments and their applications to specific problems. Participants were introduced to continuous flow FAB (fast atom bombardment) and its use for example, in the real time monitoring of biochemical reactions in vitro and in vivo; the structural elucidation of secondary metabolites from fungi; the analysis of molecule-receptor interactions; the determination of posttranslational modifications of peptides; and the location of S-S bridges in determining the tertiary structure of proteins, amongst other things.

All this provided an excellent insight into modern analytical research: the strategy and tactics of solving problems.

In summary, a very diverse picture of modern bio-organic analytical research was presented to the participants and is now collected into a permanent record which is hoped will promote further dissemination and development of this fascinating research.

Bern, 22 December, 1988 Urs Peter Schlunegger

CONTENTS

SOME CHEMICAL PHENOMENA OF
MUSHROOMS AND TOADSTOOLS

Wolfgang Steglich

Institut für Organische Chemie und Biochemie der Universität Bonn
Gerhard Domagk-Straße 1, D-53 Bonn 1, West Germany

Abstract: Mushrooms are a rich source of new natural products which often possess unusual structures and interesting biological properties.

The fruit-bodies of mushrooms and toadstools exhibit a wide variety of different colours, colour reactions, tastes, and odours which often help the mycologist to determine the species. Several mushrooms are rarely attacked by insects whereas others are easy prey to their larvae. All of these phenomena are caused by chemicals which have been studied only in recent years. In this lecture I want to draw your attention to some cases where such phenomena led to the isolation of interesting new compounds.

In the protection of fruit-bodies against insects, bitter or pungent substances play an important role. Thus the agarics *Cortinarius crystallinus*, *C. croceo-coeruleus* and *C. vibratilis* produce the strongly bitter crystallopicrin **4** (1), a triterpenoid which is concentrated in the slimy outer layer of the cap. Its structure has been determined by 2D-INADEQUATE experiments and resembles that of the iridals from *Iris* species(2). The carbon skeleton of **4** can be

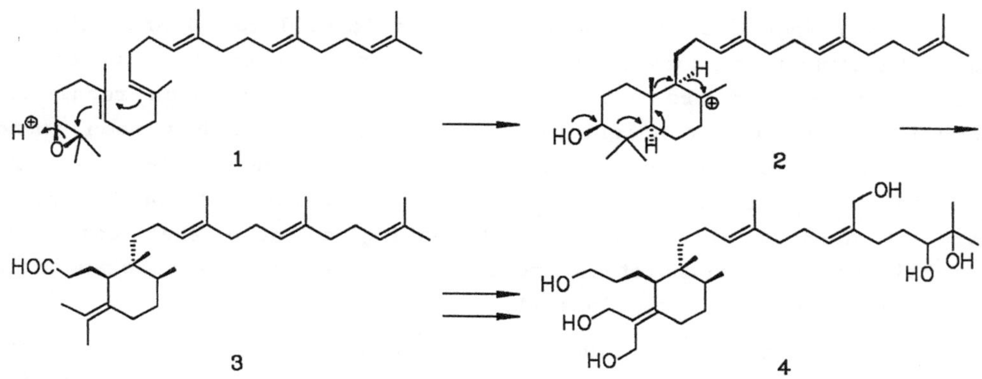

1 2 3 4

U.P. Schlunegger (Ed.)
Biologically Active Molecules

derived from squalene epoxide **1** by a shortened version of the lanosterol biosynthesis. Cyclization of **1** leads to a carbenium ion **2** which then undergoes a backbone rearrangement terminating in a fragmentation to aldehyde **3**. Reduction of **3** and consecutive hydroxylations at the allylic positions and the terminal double bond then leads to **4**. Bitter terpenoids are also found in sporophores of *Hypholoma fasciculare*(3) and *Gymnopilus spectabilis*(4) whereas *Cortinarius infractus*(5) and *Clavariadelphus pistillaris*(6) owe their bitter taste to indole alkaloids and a spermidine derivative, respectively.

Pungent compounds are even more potent antifeedants. Well known examples are the sharp tasting *Lactarius* and *Russula* species which have developed an interesting chemical defence system(7). The white milk of *Lactarius vellereus* contains the mild tasting sesquiter-penoid stearylvelutinal **5** (8) which is apparently devoid of any major biological activity. On damage to the fruit-body an enzyme is released which hydrolyses the ester function of **5** setting the stage

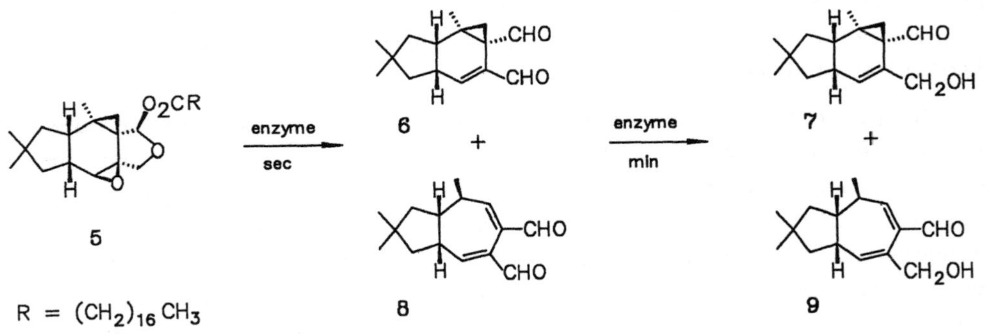

for a rearrangement into the dialdehydes isovelleral **6** and velleral **8**. Both compounds possess strong antibiotic, cytotoxic and mutagenic activities(9) and are efficient antifeedants which even repel the American opossum, a notorious mycophagist(10). It has been found that the toadstool detoxifies its own defence substances by enzymatic reduction into the corresponding alcohols **7** and **9**, which are of mild taste(7). It is interesting to note that compounds containing ene-dial moieties are often used as defence substances in the animal and plant kingdoms(11).

A completely different class of compounds is responsible for the peppery taste of the small bolete *Chalciporus piperatus*. It contains the pungent 2*H*-azepines chalciporone **10** and chalciporyl propionate **11** in admixture with the mild tasting 3*H*-azepines isochalciporone **12** and dehydroisochalciporone **13** (12). All of these substances exhibit antibiotic activities against fungi and bacteria(13). 2*H*-Azepines are a little-known class of compounds which has not been found in nature before.

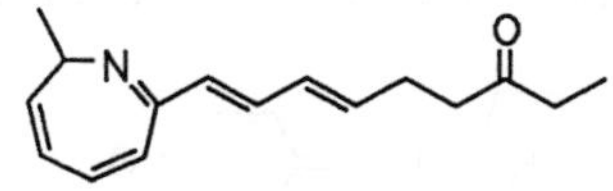

10

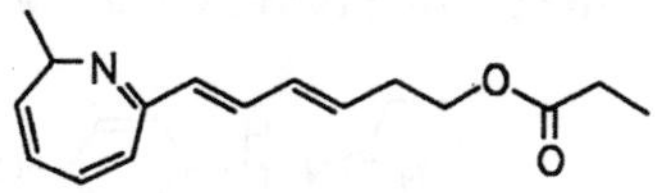

11

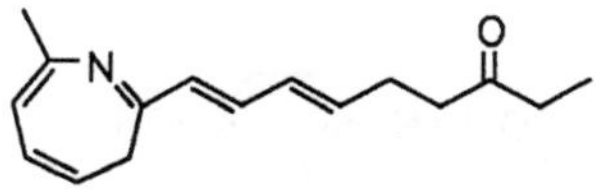

12

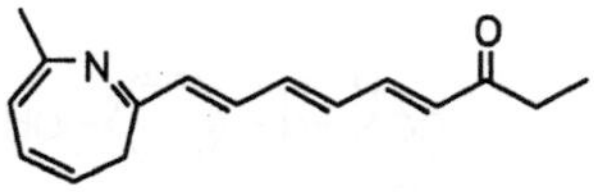

13

The defence system of *Lactarius* indicates that toadstools are able to store precursors of aggressive compounds which may be released enzymatically when needed for the protection of the fruit-bodies. Such a system is apparently also present in *Paxillus atrotomentosus*, a toadstool which is rarely infested by insects. Its fruit-bodies contain the triester leucomentin-3 **14** which on mild hydrolysis releases its epoxysorbic acid residues yielding a mixture of (+)-osmundalactone **15** and butenolide **16** (14). (-)-Osmundalactone is known as an antifeedant from leaves of the Japanese fern *Osmunda japonica*(15).

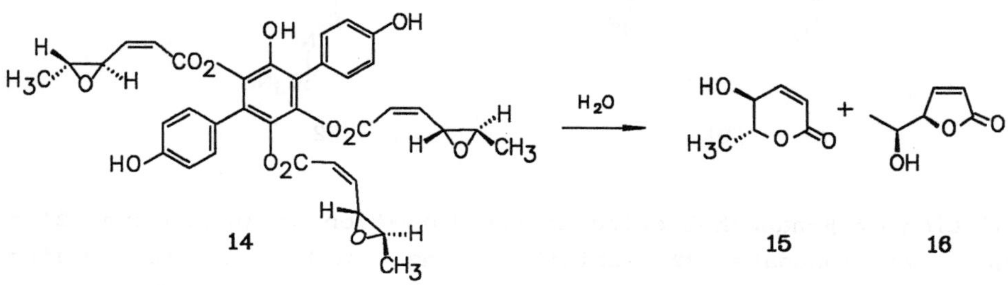

14 15 16

Another kind of chemical defence has been invented by *Agaricus xanthoderma*. This toadstool develops an intense yellow colour on bruising and is easily recognised by its carbolic smell. The methanolic extract of the fruit-bodies contains a light sensitive antibiotic which exhibits strong activity against fungi and bacteria(16). In a first investigation of *Agaricus xanthoderma*, phenol, hydroquinone, 4,4'-dihydroxydiphenyl and the pale yellow 4,4-dihydroxyazobenzene were isolated from its methanolic extract(17). Later it was shown that the colour reaction of the fruit-bodies is caused by two closely related chromogens, leucoagaricone **17** and xanthodermin **19** (18). On oxidation **17** is converted into the yellow pigment agaricone **18** whereas **19** yields the acyl diazene derivative **20** which gives rise to a yellow anion on

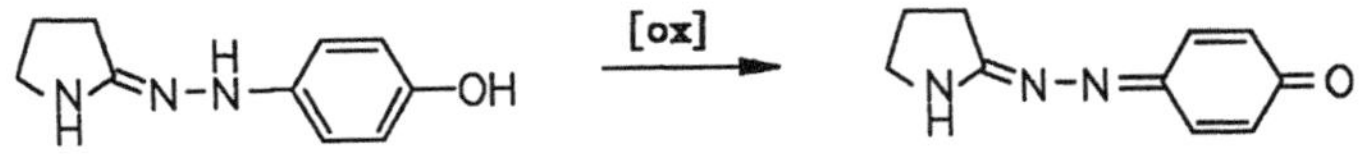

17 18

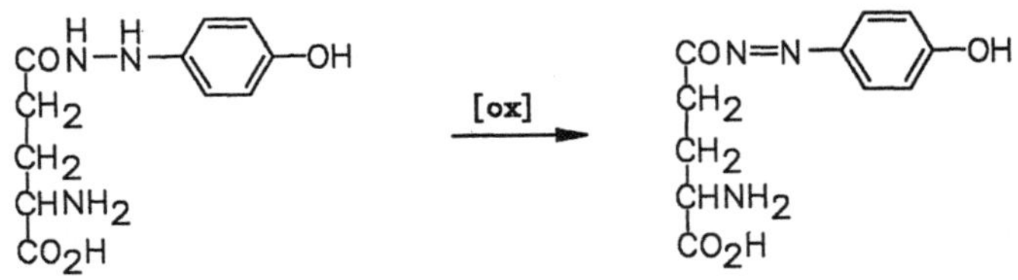

19 20

dissociation. Finally, it was found that methanolic solutions of *Agaricus xanthoderma* contain 4-diazo-2,5-cyclohexa-dien-1-one **21** which is easily recognised by the development of a red colour on

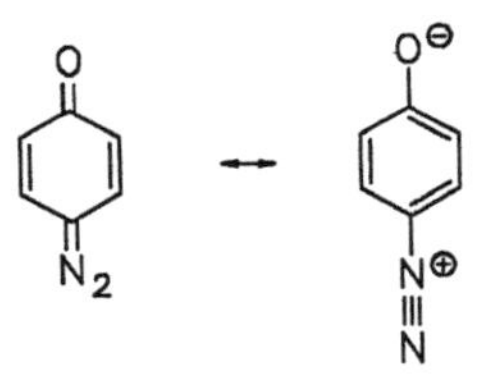

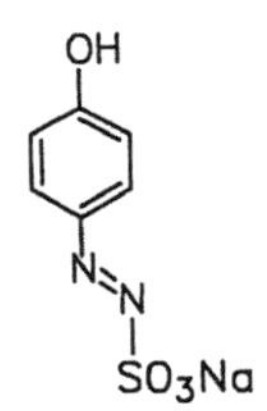

21 22

addition of β-naphthol owing to the formation of an azo dye. **21** and the diazosulphonate **22** exhibit strong antibiotic and antitumor activities(18,19). It seems possible that xanthodermin **19** acts as a precursor for **21** in the toadstool. The latter is then transformed into phenol and the aromatic compounds mentioned earlier.

Agarics have a high tendency for the formation of N-N bonds. A well known example is agaritin **23** (20) which occurs as a constituent of the cultured mushroom *Agaricus bisporus* and related species. Biosynthetically related are the aldehyde **24** and the formazane derivative **25** which have been isolated from *A. silvicola*(21).

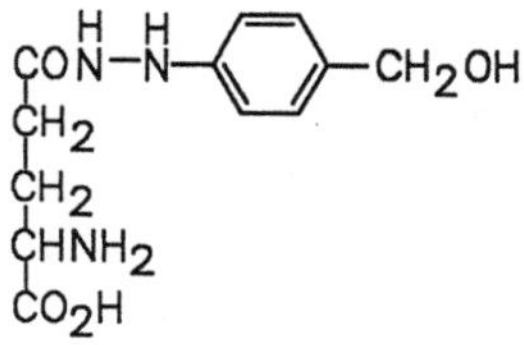

23

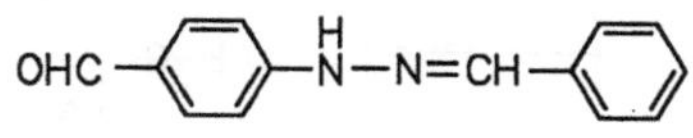

24

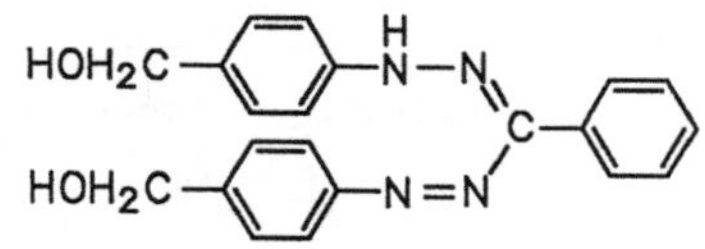

25

In some cases such compounds are health hazards like gyromitrin **26** (22), the toxin of *Gyromitra esculenta* which has caused many fatalities due to liver damage. Because **26** is easily converted into the potent carcinogen methylhydrazine, it has also to be considered as a long-time risk for human consumption(23). Potentially mutagenic substances are the nitroamines **27** and **28** which occur in *Agaricus*

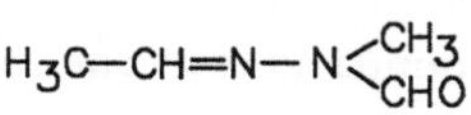

26

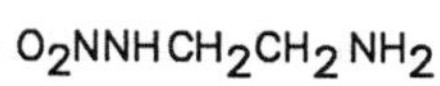

27

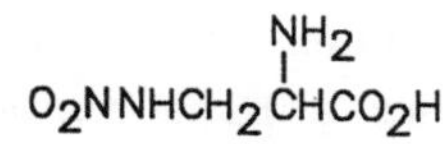

28

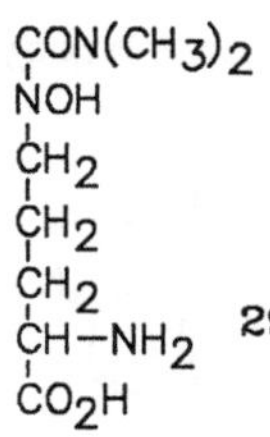

29

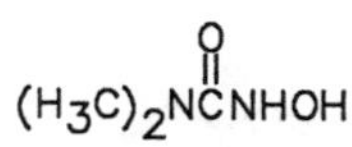

30

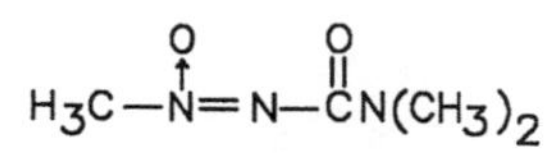

31

silvaticus(24). *Lyophyllum connatum* contains large amounts of the *N*-hydroxyornithine derivative connatin **29** together with *N*-hydroxy-*N'*,*N'*-dimethylurea **30** and the azoxy compound lyophyllin **31** (25). **29** and **30** are responsible for the intense violet colour reaction on touching the fruit-bodies with aqueous ferric chloride solution. Because these substances remain unchanged in boiling water the mushroom should be excluded from the group of edible ones.

Colour reactions are often useful as indicators for interesting metabolites. Thus the white North American puffball *Calvatia rubroflava* contains the semicarbazide derivative leucorubroflavin **32** which is immediately oxidized to the orange-red pigment rubroflavin **33** on touching the fruit-bodies(26). **33** exhibits an unusually high optical rotation due to the conjugation of the chiral sulphoxide group with the 1,4-benzoquinone chromophore.

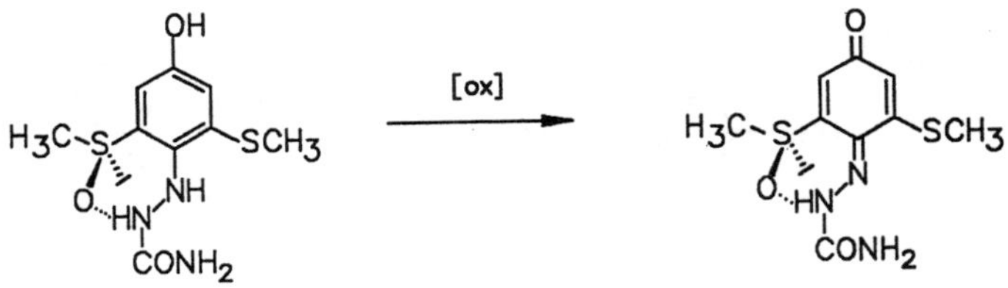

32 33

Another example for the detection of a novel type of compound by a colour reaction is *Lactarius necator*. This common toadstool changes its colour from olive-brown to purple when exposed to ammonia vapours. It could be shown that this colour change is caused by necatorone **34** an alkaloid which occurs in the skin of this toadstool together with its 4,4'-dehydrodimer **35** (27). On addition of base **34** forms a blue monoanion and a violet dianion depending on the amount of base added. From the North American relative *Lactarius atroviridis*, a toadstool of dark-green appearance, the green dimer **36** has been isolated as the main alkaloid(28). For the biosynthesis

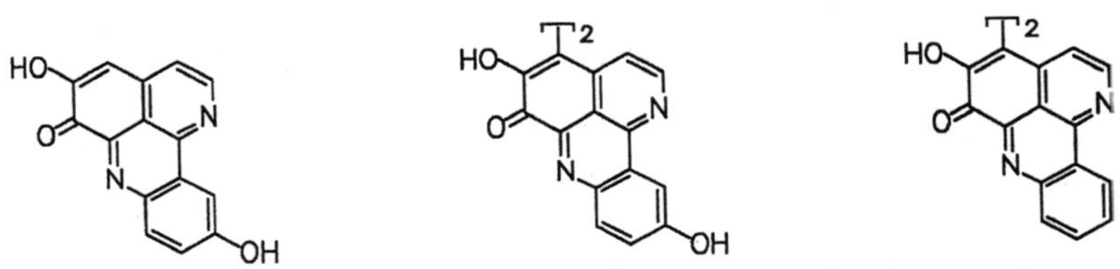

34 35 36

of necatorone and deoxynecatorone a condensation of 3,4-dihydroxy-β-phenylethylamine **37** with anthranilic acid **38** can be proposed, followed by several stages as shown in Scheme 1. In preliminary feeding experiments with ^{13}C-labeled tyrosine and anthranilic acid a high incorporation of these precursors into necatorone was observed.

Necatorone is of interest because of its strongly mutagenic activity in the Ames test which matches that of aflatoxin B_1(29). For the study of structure-activity relationships, syntheses of necatorone, deoxynecatorone and its analogues have been developed(30).

With this short selection of chemical phenomena I wanted to illustrate the importance of mushrooms and toadstools for the discovery of new biologically active natural products. Often these molecules exhibit unprecedented structures which usually appear less complex than those of the metabolites from streptomycetes and marine organisms. This may reflect the short period of time in which the biosynthesis of mushroom compounds has to take place during the development of the fruit-bodies.

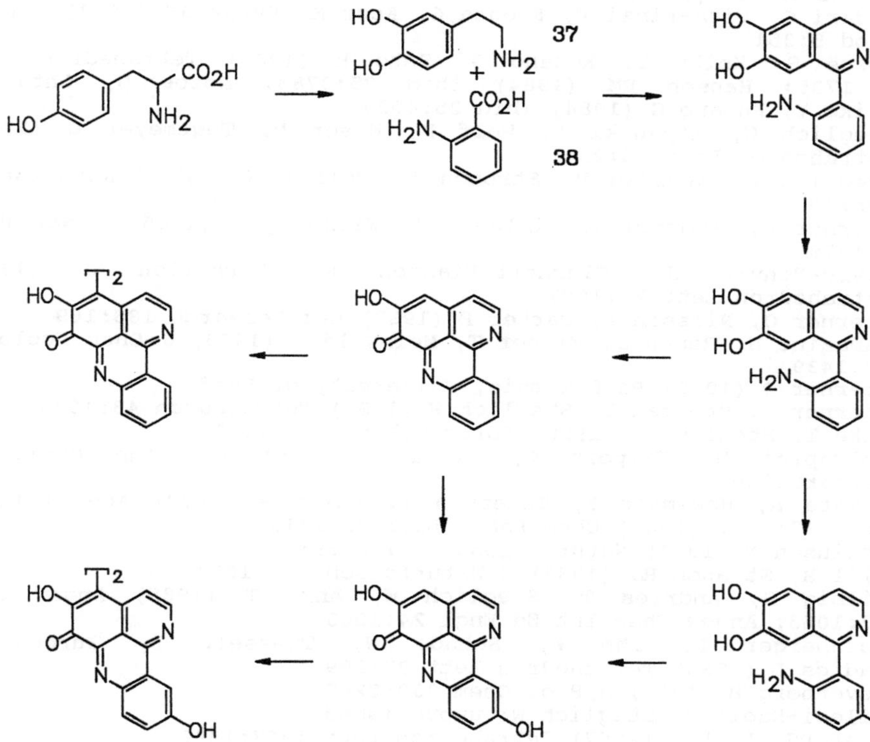

Scheme 1

Acknowledgement: I owe sincere thanks to my co-workers which carried out these investigations with great dedication and skill. The research was supported by the Deutsche Forschungsgemeinschaft and the Ministry of Science and Technology.

REFERENCES

1 Fugmann B, Koch B, Steffan B, Steglich W: in preparation.
2 Jaenicke L, Marner FJ (1986) Progr Chem Org Nat Prod 50:1
3 Ikeda M, Sato Y, Izawa M, Sassa T, Miura Y (1977) Agric Biol Chem 41:1539; Ikeda M, Watanabe H, Hayakawa A, Sato K, Sassa T, Miura Y (1977) ibid 41:1543; Ikeda M, Niwa G, Tohyama K, Sassa T, Miura Y (1977) ibid 41:1803; de Bernardi M, Mellerio G, Vidari G, Vita-Finzi P, Fronza G, Kocr M, Pyrek JS (1981) J Nat Prod 5:351
4 Nozoe S, Koike Y, Kusano G, Seto H (1983) Tetrahedron Lett 24:1735; Hanson RM (1984) ibid 25:3783; Nozoe S, Ohta T, Koike Y, Kusano G (1984) ibid 25:4023
5 Steglich W, Kopanski L, Wolf M, Moser M, Tegtmeyer G (1984) Tetrahedron Lett 25:2341
6 Steglich W, Steffan B, Stroech K, Wolf M (1984) Z Naturforsch 39c:10
7 Sterner O, Bergman R, Kihlberg J, Wickberg B (1985) J Nat Prod 48:279
8 Favre-Bonvin J, Gluchoff-Fiasson K, Bernillon J (1982) Tetrahedron Lett 23:1907
9 Sterner O, Nilsson L, Carter R (1987) Mut Research 188:169
10 Camazine S, Resch J, Eisner T, Meinwald J (1983) J Chem Ecology, 10:1439
11 Sterner O (1985) Ph D Thesis, University of Lund
12 Sterner O, Steffan B, Steglich W (1987) Tetrahedron 43:1075
13 Anke T, Steglich W (1988) Forum Mikrobiol 11:21
14 Holzapfel M, Kilpert C, Steglich W Liebigs Ann Chem: in preparation
15 Numata A, Hokimoto K, Takemura T, Fukui S (1983) Appl Entomol Zool 18:129; (1984) Chem Pharm Bull 32:2815
16 Atkinson N (1954) Nature (London) 174:598
17 Gill M, Strauch RJ (1984) Z Naturforsch 39c:1027
18 Hilbig S, Andries T, Steglich W, Anke T (1985) Angew Chem 97:1063; Angew Chem Int Ed Engl 24:1063
19 Dornberger K, Ihn W, Schade W, Tresselt D, Zureck A, Radics L (1986) Tetrahedron Lett 27:559
20 Levenberg B (1964) J Biol Chem 239:2267
21 Kileci-Ksoll R, Steglich W: unpublished
22 List PH, Luft P (1967) Tetrahedron Lett 1967:1893
23 Bresinsky A, Besl H (1985): Giftpilze: Ein Handbuch fuer Apotheker, Aerzte und Biologen, Wissenschaftliche Verlagsgesellschaft mbH, Stuttgart
24 Chilton WS, Hsu CP (1975) Phytochemistry 14:2291
25 Fugmann B, Steglich W (1984) Angew Chem 96:71; Angew Chem Int Ed Engl 23:72
26 Fugmann B, Steglich W (1987) In: Gill M, Steglich W, Progr Chem Org Nat Prod 51:243
27 Fugmann B, Steffan B, Steglich W (1984) Tetrahedron Lett 25:3575
28 Klamann J-D, Fugmann B, Steglich W: Phytochemistry in press
29 Suortti T, von Wright A (1983) J Chrom 265:529; Suortti T (1984) ibid 301:303
30 Hilger CS, Fugmann B, Steglich W (1985) Tetrahedron Lett 26:5975 Hilger CS (1988) Dissertion, University of Bonn

β-METHOXYACRYLATE ANTIBIOTICS: FROM BIOLOGICAL ACTIVITY TO SYNTHETIC ANALOGUES

Timm Anke[a] and Wolfgang Steglich[b]

Abteilung Biotechnologie der Universität Kaiserslautern[a]
Paul-Ehrlich-Straße 22, D-6750 Kaiserslautern
Institut für Organische Chemie und Biochemie der Universität Bonn[b]
Gerhard-Domagk-Straße 1, D-5300 Bonn 1

Abstract: The biological activities, structures, and structure-activity relationships of strobilurins, oudemansins, and their synthetic analogues are described.

INTRODUCTION

Strobilurus tenacellus is a small agaric which grows early in the year on pine cones. In our search for new antibiotics from basidiomycetes we detected that cultures of *Strobilurus tenacellus* produce two strongly antifungal compounds which were named strobilurins A and B (1). Both antibiotics could be extracted from the mycelia with methanol and purified by chromatography on silica gel and Sephadex LH-20.

STROBILURINS AND OUDEMANSINS

The structures of strobilurin A and B were determined as **1a** and **1b** from their spectroscopic data (2,3). The compounds can be easily hydrogenated to their tetrahydro derivatives **2** which give rise to characteristic fragmentations in the mass spectrum as indicated in the formula. The substitution pattern in the aromatic ring of strobilurin B was secured by Lemieux - von Rudloff oxidation to 4-chloro-3-methoxybenzoic acid (3).

Strobilurins and similar β-methoxyacrylates can be recognised in the mass spectrum by the occurrence of an intense ion *m/z* 75 which arises by 1,3-migration of the ester methoxy group (2). In the ^{1}H-NMR spectrum the coupling pattern of the vinylic protons, the presence of two methoxy singlets and a one proton singlet at δ = 7.4 ppm are useful characters for the identification of strobilurins.

U. P. Schlunegger (Ed.)
Biologically Active Molecules

The latter signal can be assigned to the (*E*)-β-methoxyacrylate unit and occurs at higher field (Δ δ = 0.6 - 0.8) if the configuration is reversed.

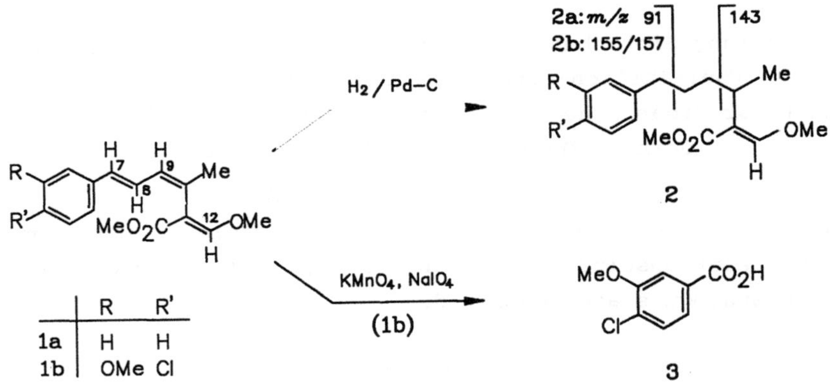

The assignment of the stereochemistry at the 9-double bond proved to be more difficult (2) and was finally established by comparison of the NMR data of strobilurin A **1a** with those of synthetic (9*E*)-strobilurin **4** (3) (Scheme 1):

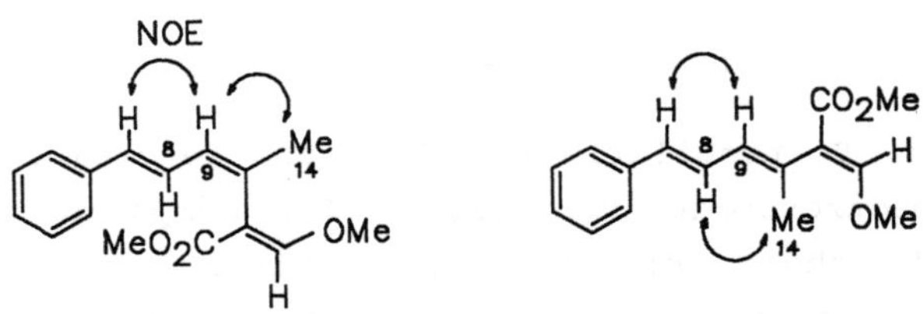

1a 4

Characteristic NMR data ($CDCl_3$)		
δ 8-H:	6.70	7.19 ppm
C-14:	23.9	17.5 ppm
$J_{C\text{-}14,H\text{-}9}$:	6.5	8.0 Hz

Scheme 1

The ^1H NMR data indicate a considerable shielding of 8-H in the natural strobilurins which can be explained by a nonplanar arrangement of the β-methoxyacrylate unit due to steric interactions with the dienyl moiety. Most indicative for the assignment of the configuration is the difference in chemical shifts of the Me-14 carbon signals. In strobilurin A **1a** and its natural analogues this signal appears at δ = 23.9, whereas in the (9*E*)-isomer **4** it is shifted to 17.5 ppm in accordance with the stronger steric compression of the methyl group (4). The larger coupling constant $J_{C-14,9-H}$ of **4** (8 Hz) as compared to **1a** (6.5 Hz) is in agreement with the *trans* coupling in the former compound. The nuclear Overhauser enhancements given in the formulas support the stereochemistry which was finally secured by a single crystal *X*-ray structure determination of (9*E*)-strobilurin **4** (5).

A second type of β-methoxyacrylate antibiotic, oudemansin A, was detected in cultures of *Oudemansiella mucida* (6). Oudemansin A is an optically active solid ($[\alpha]_D$ -17°, m.p. 44 ^oC) which exhibits a high activity against a wide variety of fungi comparable to that of the strobilurins. Its structure **5a** was determined by an *X*-ray crystal structure analysis. **5a** represents a strobilurin which has undergone

5a 6

Selected NMR data ($CDCl_3$)		
δ 7–H:	6.46	6.60 ppm
8–H:	5.92	6.02
9–H:	3.96	4.04
10–H:	2.99	3.00
14–H:	1.26	1.08

Scheme 2

methanol addition at the 9-double bond. This creates two neighbouring chiral centers which possess *syn*-stereochemistry. As can be seen from Scheme 2, **5a** and the synthetic *anti*-diastereomer **6** (7,8) show very similar ^1H NMR spectra, the occurrence of the Me-14 doublet at lower field in **5a** being the most distinctive feature. The coupling constants $J_{9-H,10-H}$ ≈ 10 Hz are nearly identical in both cases and are in accordance with the conformations indicated in formulas **5a** and **6**. Interestingly, the (11*Z*)-isomers of **5** and **6** exhibit a smaller *J*-value of 7.5 Hz (7). Apparently, in these cases the *gauche*- and *anti*-conformations, respectively, are somewhat distorted under the influence of the (*Z*)-β-methoxyacrylate unit. The absolute configuration of oudemansin A was finally established as (9*S*,10*S*) by means of an enantioselective total synthesis which will be discussed later (9).

In 1969, an antifungal antibiotic mucidin had been isolated by Musílek et al. (10) from cultures of *Oudemansiella mucida*. The compound was described as crystalline (m.p. 51-53 ^{o}C) and optically active ($[\alpha]_{546}$ +33°) (11) and the same molecular formula was given for mucidin as for strobilurin A which, however, is an oil and optically inactive. We therefore concluded that mucidin was neither identical with strobilurin A nor oudemansin A (2,6). Later, the Czechoslovakian group claimed that mucidin is identical with strobilurin A (12) which may lead to confusion in naming the compounds (13).

In a broad screening of cultures of related genera (14) we found that strobilurins and oudemansins are produced by about 20 species of basidiomycetes. Thus we could isolate compounds of this type from cultures of *Crepidotus*, *Cyphellopsis*, *Hydropus*, *Mycena*, *Oudemansiella*, *Strobilurus*, and *Xerula*. In the course of these studies we detected strobilurin C **1c** as a new member of this group

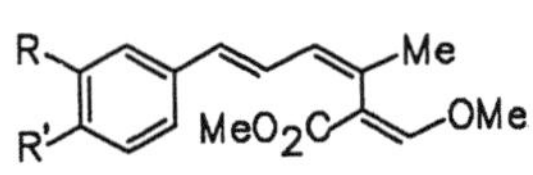

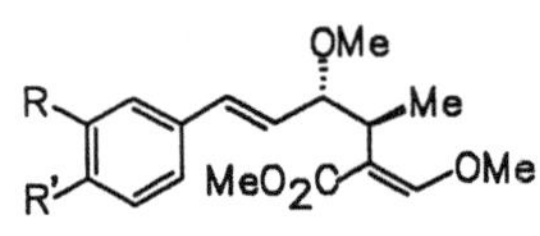

	R	R'	
1a	H	H	5a
1b	MeO	Cl	5b
1c	PrenylO	H	
1d	H	MeO	5d

in cultures of *Xerula longipes* (15). Another analogue **1d** with a *p*-methoxy substituent at the phenyl ring was isolated by Vondrácek et al. (16) from cultures of *Oudemansiella mucida*. It is interesting to note that the oudemansins B **5b** and X **5d** corresponding to strobilurins **1b** and **1d**, respectively, have been found in cultures of *Xerula melanotricha* (15) and *Oudemansiella radicata* (17) recently. The occurrence of strobilurins and oudemansins with the same substitution pattern indicates a close biosynthetic relationship.

A more elaborate strobilurin derivative is hydroxystrobilurin D **7** which occurs in cultures of *Mycena sanguinolenta* (18). Its structure was determined from the spectroscopic data and the conversion into a

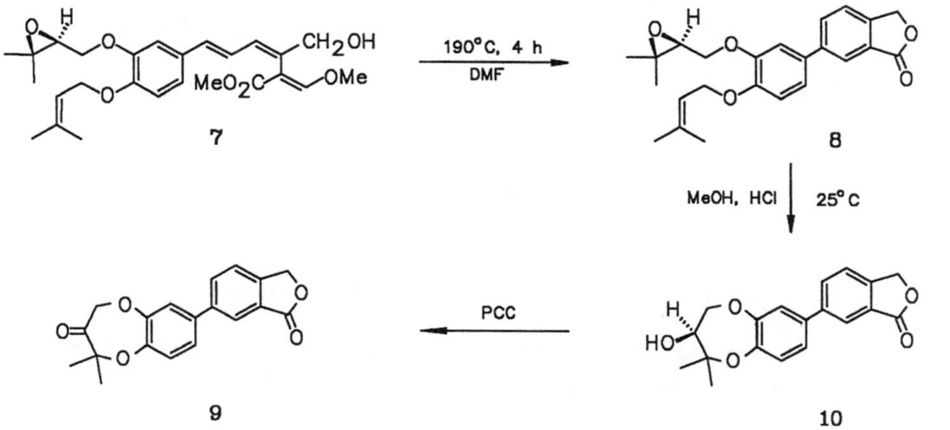

diphenyl derivative **8** on heating in DMF at 180 °C. The thermolysis induces an electrocyclic ring closure of the triene unit followed by elimination of methanol and formation of the phthalide ring by intramolecular transesterification. The same type of aromatization is observed with strobilurin A. However, in this case a higher temperature is required (3). Interestingly, the cyclization does not occur with (9*E*)-strobilurin **4** because the presence of the (9*E*)-double bond precludes the geometry of the transition state necessary for cyclization. The biphenyl derivative **8** undergoes loss of the prenyl residue and ring closure to a 1,4-dioxepine derivative **10** under mildly acidic conditions. The presence of the seven-membered ring was secured by oxidation of **10** to ketone **9**. The alcohol **10** was used to establish the (*S*)-configuration for hydroxystrobilurin D by means of Horeau's method.

The biosynthesis of strobilurin A has been studied (19) and it has been found that phenylalanine and benzoic acid are efficiently incorporated into the phenyl ring and C-7 (Scheme 3). Feeding of labeled methionine revealed that not only the methoxy groups but also Me-14 are derived from this precursor. By use of [1-^{13}C]- and [2-^{13}C]-labeled acetate the origin of carbons C-8 to C-12 of the side chain could be clarified. Interestingly, the ester carbon C-13 is derived from C-1 of acetate which suggests a 1,2-shift of the terminal ester group in a hypothetical polyketide precursor.

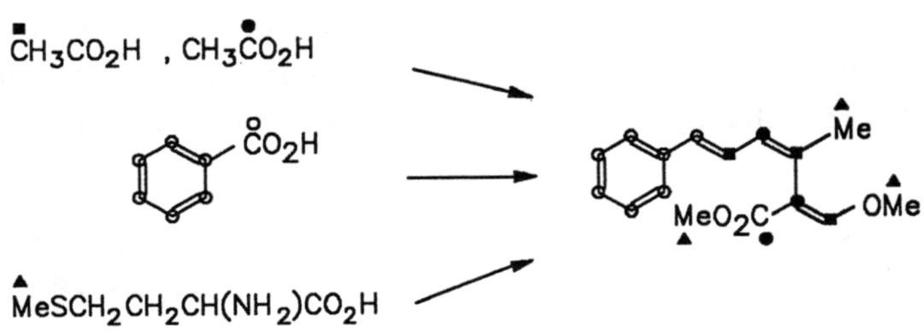

Scheme 3

SYNTHESIS OF NATURAL β-METHOXYACRYLATES

Strobilurin A **1a** and (9*E*)-strobilurin A **4** and have been prepared by a stereocontrolled synthesis as shown in Scheme 4 (3).

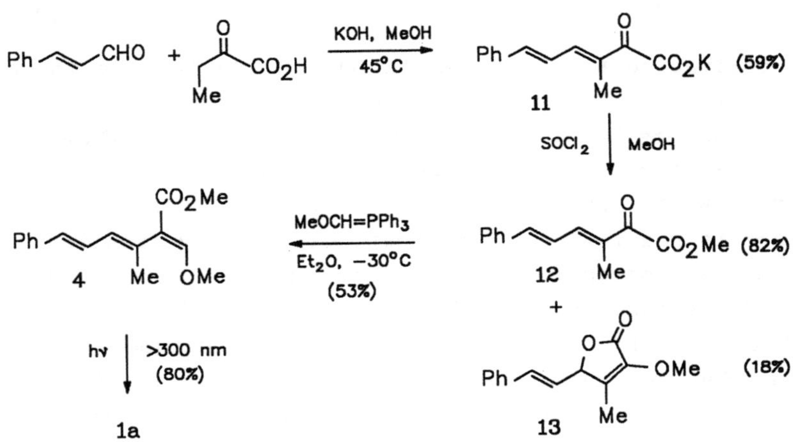

Scheme 4

The synthesis starts from α-oxo acid **11** whose configuration was established by oxidative degradation to (2*E*,4*E*)-5-phenylpentadienoic acid of known stereochemistry. Esterification of **11** with thionyl chloride/methanol leads to a mixture of the desired ester **12** and a cyclization product **13** which can be directly converted into crystalline **4** by treatment with (methoxymethylene)triphenylphosphorane. Irradiation of **4** with a high-pressure mercury lamp at wavelengths > 300 nm in benzene containing some acetone leads to an inversion of the 9-double bond. The strobilurin A **1a** formed was identical with the natural compound.

Recently, a second stereoselelective synthesis of strobilurin A has been developed (20). The synthesis starts from the (2*Z*,4*E*)-pentadienoic ester **14** which was converted in five steps into the α-oxo ester **15**. The usual Wittig procedure then yielded strobilurin A.

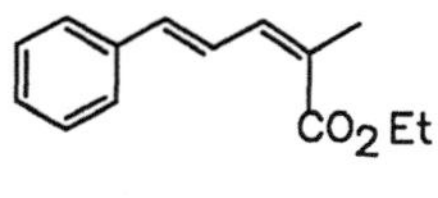

14

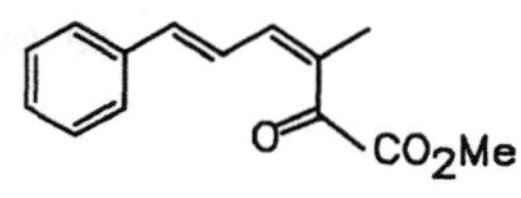

15

The first synthesis of (±)-oudemansin A was carried out by Oishi and co-workers (21). This synthesis has been considerably improved by a novel method for ester homologation (Scheme 5) (22).

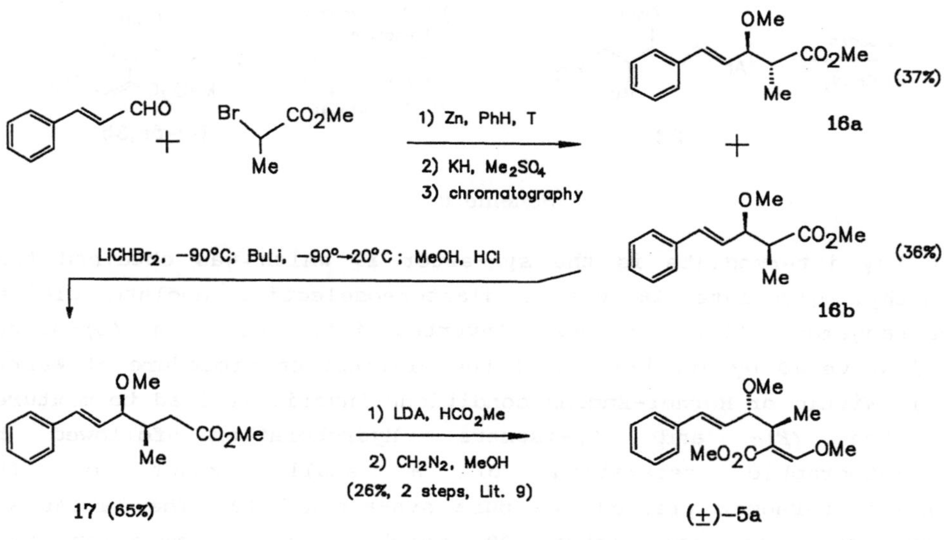

Scheme 5

Reformatsky reaction of cinnamaldehyde and methyl α-bromo propionate yielded a 1:1 mixture of the *syn*- and *anti*-hydroxy esters which was directly O-methylated. Chromatographic separation of this mixture gave the desired *syn*-diastereomer **16b** and the *anti*-compound **16a** in nearly equal amounts. The chain elongation was carried out by treatment of **16b** with dibromomethyllithium at -90 °C followed by addition of butyllithium and warming up to 30 °C. After quenching with acidic methanol the desired homologous ester **17** was obtained in 65% yield without any loss of its sterical integrity (22). Formylation of **17** and subsequent treatment with diazomethane yielded **5a** in moderate yield (21).

A flexible synthesis of (±)-oudemansins was developed by Wittman and Kallmerten (23) (Scheme 6).

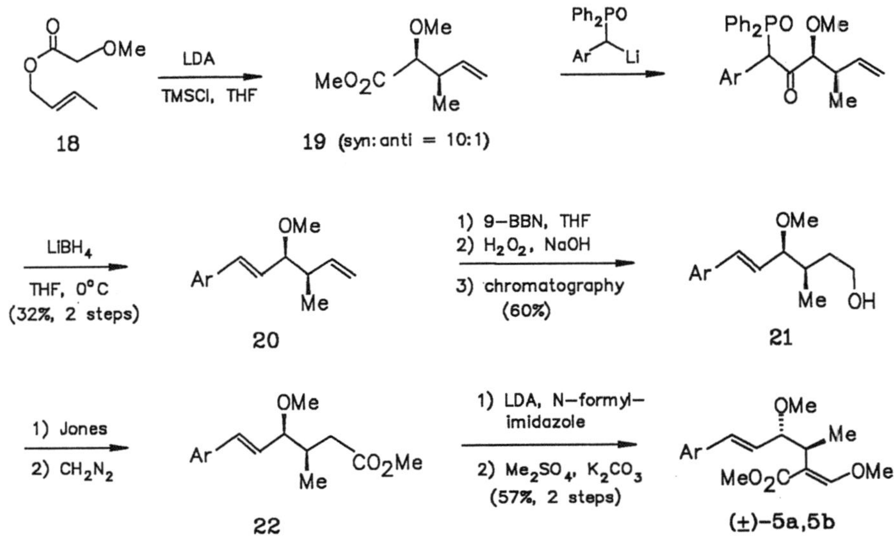

Scheme 6

The key intermediate is the *syn*-ester **19** which was obtained from *O*-methyl glycolate **18** *via* a diastereoselective enolate Claisen rearrangement (24). **19** was converted into the pure (*E*)-styryl derivative **20** by utilization of the olefination procedure of Warren (25). Wittig or Horner-Emmons conditions invariably lead to mixtures of the (*E*)- and (*Z*)-isomers. Hydroboration followed by chromatographic separation of a small amount of the *anti*-diastereomer yielded the pure *syn*-alcohol **21**. The latter was converted into the ester **22** which after formylation with *N*-formylimidazole and subsequent methylation yielded (±)-oudemansin A **5a**. In a similar fashion (±)-oudemansin B **5b** was obtained from **19**.

Condensation of the trimethylsilyl enol ether **23** with cinnamaldehyde dimethyl acetal **24** under Mukaiyama conditions followed by reaction with (methoxymethylene)triphenylphosphorane yields (±)-(11*Z*)-epioudemansin A **25** as the main product in admixture with smaller amounts of epioudemansin A **6**, (11*Z*)-oudemansin A, and traces of oudemansin A (7) (Scheme 7). The stereoisomers can be separated by chromatography on silica gel. Irradiation of the (11*Z*)-isomers with UV light at wavelenghts >300 nm in benzene/acetone solution leads to inversion of the configuration at the β-methoxyacrylate unit.

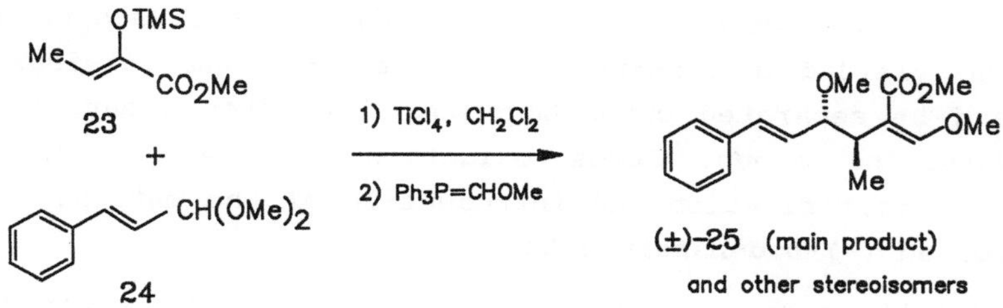

Scheme 7

The first synthesis of (-)-oudemansin A was achieved by Oishi and co-workers (9) (Scheme 8).

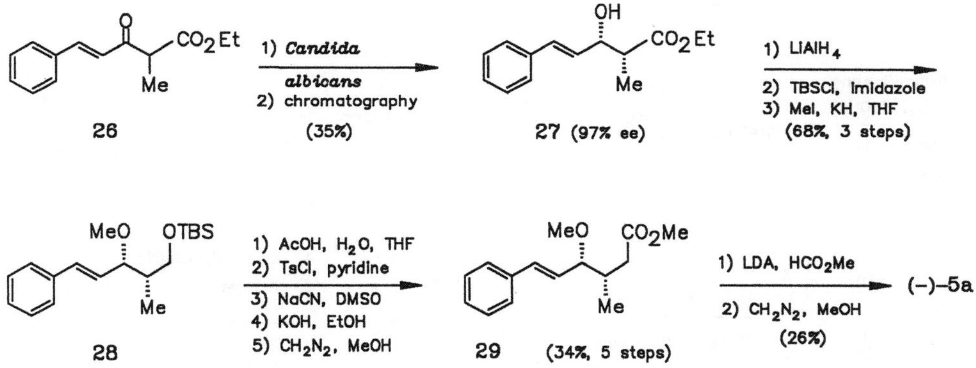

Scheme 8

It starts from β-oxo ester **26** which on microbial reduction with *Candida albicans* afforded the optically active (97% ee) *syn*-β-hydroxy ester **27** in 35% yield along with 10% of the corresponding *anti*-isomer and some starting material. After chromatographic purification **27** was reduced with $LiAlH_4$. The resulting diol was selectively silyl protected at the primary hydroxy group followed by O-methylation to **28** and conversion into the homologous ester **29** by standard methods. After introduction of the methoxymethylene residue (-)-oudemansin **5a** was obtained which was identical with the natural product. This proves its absolute configuration as 9*S*,10*S*.

The (9*S*,10*S*)-configuration of oudemansin B was established in a similar fashion (26) (Scheme 9). The crucial intermediate in this synthesis is *syn*-aldehyde **31** which was generated from the optically pure epoxy ester **30** by the steps indicated in the Scheme. Wittig reaction of **31** with (4-chloro-3-methoxybenzylidene)triphenyl-phosphorane yielded a 1:1-mixture of the (*E*)- and (*Z*)-isomers **32** which could be separated chromatographically after cleavage of the silyl protecting group. Jones oxidation of the (*E*)-alcohol **33** followed by esterification and introduction of the methoxymethylene group afforded (-)-oudemansin B **5b**.

Enantiomerically pure stereoisomers of the β-hydroxy ester **27** were also obtained by a different route and served as starting materials for syntheses of (+)-oudemansin A and (+)-epioudemansin A (8).

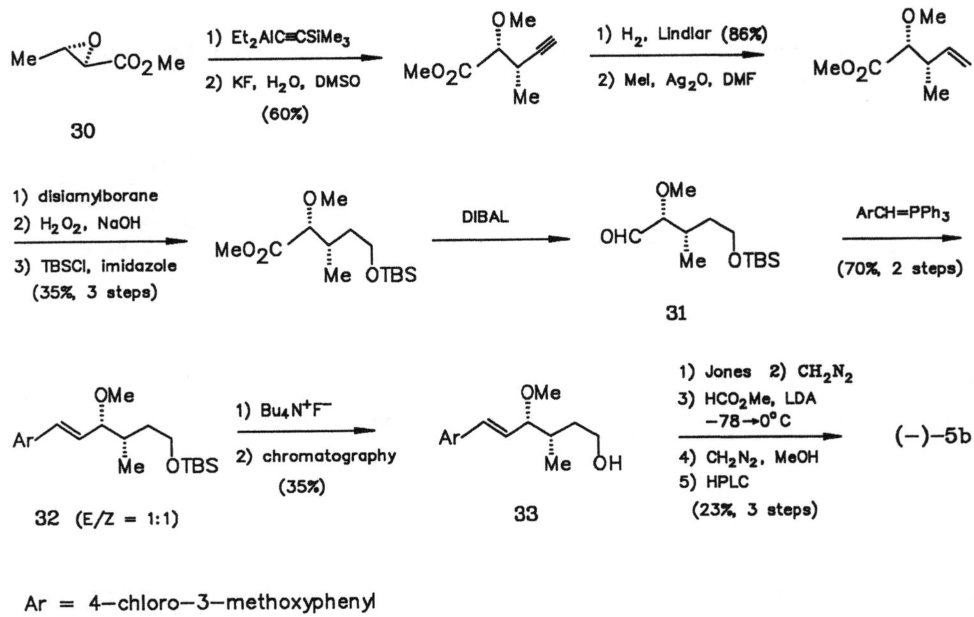

Scheme 9

BIOLOGICAL ACTIVITIES OF STROBILURINS AND OUDEMANSINS

The evaluation of the biological activities of the strobilurins and oudemansins revealed a very strong antifungal activity against yeasts and filamentous fungi at very low concentrations starting from below 1 µg/ml (Scheme 10) (1). No antibacterial activity could be detected. The inhibition of fungal growth was most pronounced when the fungi were grown under conditions favoring aerobic metabolism. A complete inhibition of the oxygen uptake by freshly germinated spores of *Penicillium notatum* was observed at concentrations of 0,2 µg/ml (Scheme 10). In rat liver mitochondria respiration and ATP synthesis with both α-ketoglutarate and succinate were completely inhibited by concentrations of 10^{-6} M strobilurin A (6). Similar results have been obtained by Subík et al. (27).

The inhibition of eucaryotic respiration by strobilurins and oudemansins is specific and reversible. Cells of the ascitic form of Ehrlich carcinoma (mouse) can derive metabolic energy from both anaerobic glycolysis and oxidative phosphorylation. Only in the absence of free glucose in the media the incorporation of precursors into protein, RNA, and DNA is completely inhibited by strobilurins

strobilurin A (1a)	strobilurin B (1b)	strobilurin C (1c)
10/40[1] 1/30 100%/0.2[2]	10/40 1/20 100%/0.2	10/22 1/16 100%/0.2
hydroxystrobilurin D (7)	**strobilurin E (55)**	**(9*E*)-strobilurin A (4)**
10/40 1/21 100%/3-5	5/15 93%/0.04	10/25 100%/17
oudemansin A (5a)	**oudemansin B (5b)**	**oudemansin X (5d)**
10/40 1/20 100%/0.2	10/40 1/22 100%/0.2	10/38 1/30 75%/1

Scheme 10. Antibiotic Activity of Strobilurins and Oudemansins

1) Plate diffusion assay, ***Penicillium notatum***, µg/disc vs. diameter inhibition zone (mm)

2) Inhibition of respiration, *P. n.*, % inhibition of oxygen uptake vs. concentration (µg/ml)

and oudemansins. At a dosage of 9 times 36 mg/kg i. p. strobilurin A prolonged the life span of mice with lymphocytic leucemia by 20 % as compared to the controls receiving no treatment (28). The very low toxicity *in vivo* seemed to be quite exceptional for an inhibitor of mitochondrial respiration.

The precise target of the strobilurins and oudemansins within the mitochondrial electron transport chain was located by von Jagow et al. (29,30). The antibiotics specifically inhibit the Q_O reaction center of the bc_1 complex. A shift of the relative position of the ubihydroquinone molecule in the reaction centre, due to a conformational distortion of cytochrome *b* induced by the binding of the (*E*)-β-methoxyacrylate (MOA) antibiotic, seems to be the reason for the blocked electron transfer (31).

SYNTHESIS OF ANALOGUES, STRUCTURE-ACTIVITY RELATIONSHIPS

The unique structures of the strobilurins and oudemansins in combination with their selective antifungal activity induced us to synthesize simpler analogues to study their structure-activity relationships.

The investigation of the antifungal and respiration-inhibiting activities of the natural products and their synthetic analogues depicted in Scheme 11 led to the following conclusions (32):

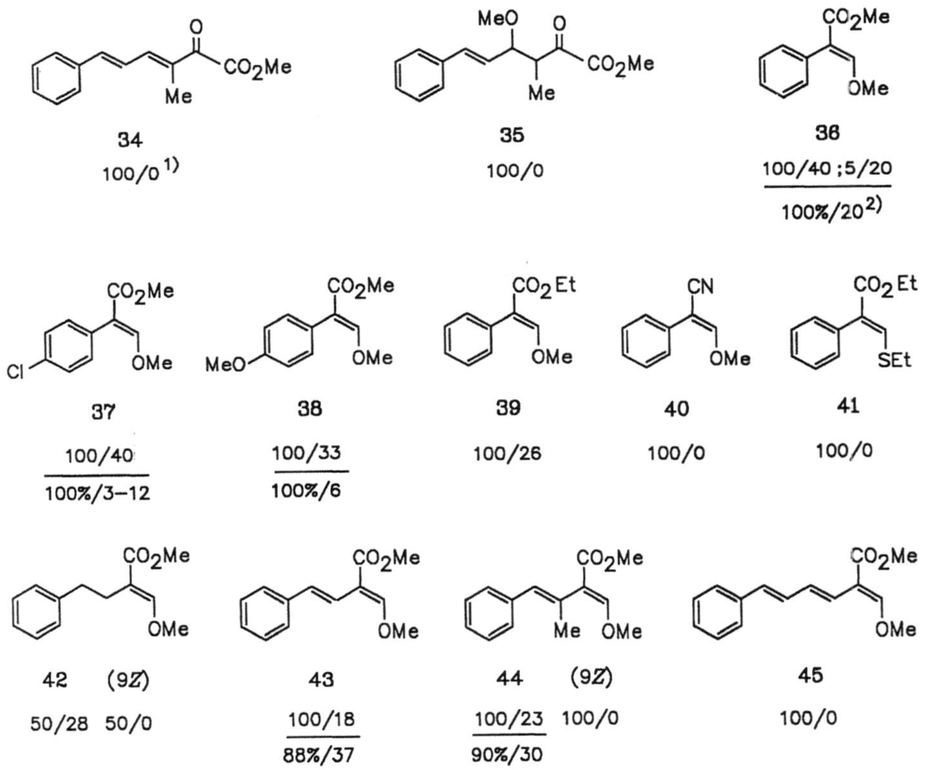

Scheme 11. Antibiotic Properties of Simple Strobilurin Analogues

1) Plate diffusion assay, *Penicillium notatum*, µg/disc vs. diameter inhibition zone (mm)

2) Inhibition of respiration, *P. n.* , % inhibition of oxygen uptake vs. concentration (ug/mm)

1. The terminal (*E*)-β-methoxyacrylate unit is a prerequisite for the antifungal and respiration inhibiting acitivities. The α-oxoesters **34** and **35** corresponding to strobilurin A and oudemansin A respectively, are inactive. The replacement of the carboxymethyl group in **36** by a cyano group (**40**) or the change in the (*E*)-configuration of the terminal methoxyacrylate unit in **42** and **44** to the (*Z*)-form leads to a complete loss of antifungal activity. Replacement of the enol ether by a thioether in **41** yields an antibacterial antibiotic.

2. The nature of the substituent at the β-methoxyacrylate unit has a marked impact on the antibiotic activity. The simple derivative **36** which lacks the central diene unit of strobilurin A exhibits a modest but selective antifungal and respiration inhibiting activity. The *p*-chloro and *p*-methoxyderivatives **37** and **38** are somewhat better antifungals. A comparison of the antifungal activities of **45** (none), (9*E*)-strobilurin A **4** (small), and strobilurin A (high) proves the decisive role of the configuration at the 9-double bond for the antifungal activity. This marked influence of the molecular geometry is also observed in the oudemansin series: Epioudemansin A **6**, differing from oudemansin A **5a** in its linear shape, is completely inactive.

These results indicate that synthetic analogues with higher activities have to be closer mimics of the natural products. A strikingly simple solution of this problem was found by incorporation of a phenyl ring into the strobilurin side chain. This change should lead to a fixation of the crucial geometry at the 9-double bond and a stabilization of the triene system:

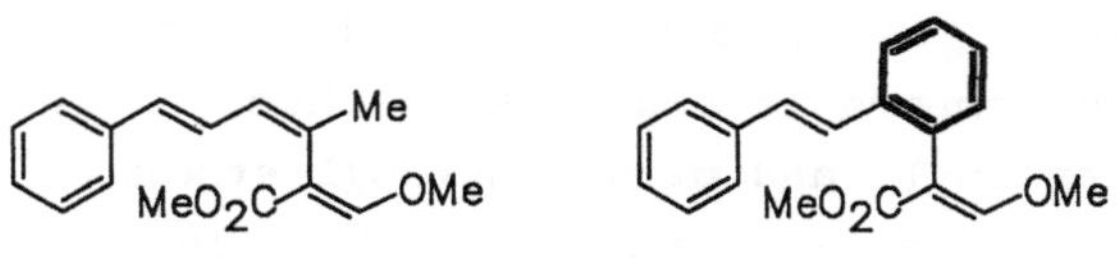

Compounds of this type are easily accessible by synthesis (33). Thus, **46** can be prepared in high overall yield from the simple β-methoxyacrylate **47** by bromination with NBS followed by Arbuzov reaction and Horner-Emmons olefination of the resulting phosphonate ester **48** with benzaldehyde (Scheme 12):

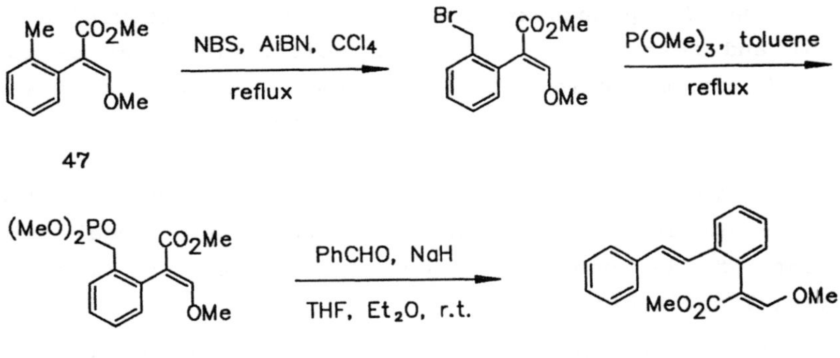

Scheme 12

The biological activities of **46** and some of its analogues are summarized in Scheme 13. The highest antifungal and respiration inhibiting activities are exhibited by the parent compound **46**. The activities of the diene homologue **49** are lower. Hydrogenation of stilbene **46** to its dihydro derivative **50** causes a strong decrease in the antifungal and respiration inhibiting activities. The tetrahydro derivative of diene **49** is completely inactive. The ether **51** and thioether **52** are both good antifungals and respiration inhibitors. Oxidation of **52** to the corresponding sulfoxide or sulfone leads to reduction or loss of the antibiotic activity, respectively. Methyl 2-(2-hydroxyphenyl)-3-methoxypropenoate (**53**) itself exhibits only very weak antifungal activities but gives rise to potent derivatives. The antifungal and respiration inhibiting activity of the benzyl ether **54** is high and comparable to that of its isomer **51**. From a series of aliphatic ethers, the geranyl ether **55** exhibits the highest activities which compare favorable with those of the strobilurins. The cinnamyl, farnesyl, and geranylgeranyl ethers of **53** as well as their hydrogenation products are less potent antibiotics.

The potential of stilbene **46** and its analogues for practical applications as fungicides is under active investigation (34).

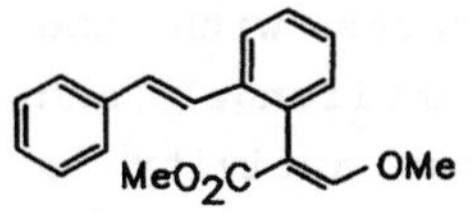

46

1/25[1]
0.1/20
———
90%/0.02[2]

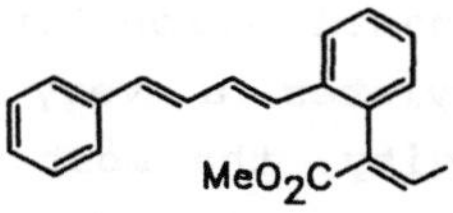

49

10/18
1/12
———
91%/0.5

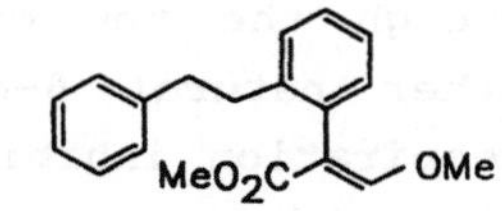

50

10/20
1/15
———
84%/10

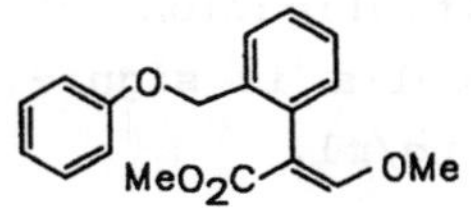

51

10/28
1/20
———
95%/0.5

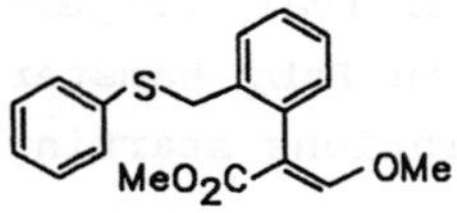

52

10/20
1/17
———
92%/0.5

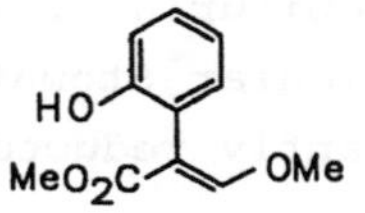

53

100/14
———
18%/100

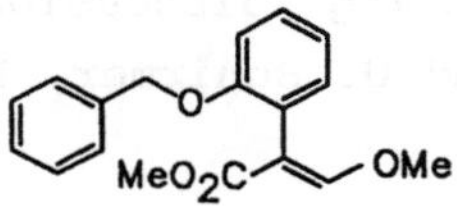

54

10/13
1/12
———
93%/1

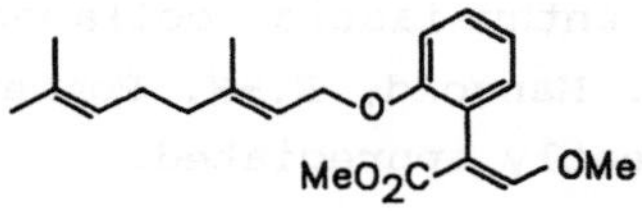

55

1/12
0.1/10
———
90%/0.1

Scheme 13. Advanced Synthetic Strobilurin Analogues

1) Plate diffusion assay, *Penicillium notatum*, µg/disk vs. diameter inhibition zone (mm)
2) Inhibition of respiration, *P. n.*, % inhibition of oxygen uptake vs. concentration (µg/ml)

FUTURE PERSPECTIVES

New interesting and potentially useful activities of β-methoxyacrylates have been discovered recently. A screening for new metabolites capable of reversibly inducing morphological alterations of tumor cells resulted in the isolation of a new strobilurin from cultures of the agaric *Crepidotus fulvotomentosus* (35). This fungus does not belong to the family *Tricholomataceae* as do most of the other producers of strobilurins and oudemansins.

Though the new compound, named strobilurin E, shares with the other natural β-methoxyacrylates a very strong antifungal and respiration inhibiting activity, the most conspicuous activity is its inhibitory action on the growth of human tumor (HeLa) cells starting at concentrations of 2 ng/ml. From concentrations of 25 ng/ml of strobilurin E, HeLa S3 cells show interesting alterations from epithelial to fibroblast-like morphology. These alterations and the cytostatic effects can be completely reversed after removal of the antibiotic from the cells media.
Strobilurin E exhibits antiviral properties. Multiplication of Vesicular Stomatitis Virus in Baby Hamster Kidney cells is significantly reduced by concentrations starting from 25 ng/ml.

Acknowledgement - Financial support by the Deutsche Forschungsgemeinschaft, the Bundesministerium für Forschung und Technologie and the BASF AG, Ludwigshafen, is gratefully acknowledged. We thank our co-workers S. Backens, J. Bäuerle, U. Mocek, G. Schramm, A. Scherer, B. Schwalge, B. Steffan, and W. Weber for their enthusiastic collaboration. Stimulating discussions with Drs. D. Mangold, E.-H. Pommer, H. Sauter and U. Schirmer, BASF AG, are greatly appreciated.

REFERENCES

1 Anke T, Oberwinkler F, Steglich W, Schramm G (1977) J Antibiot 30:806

2 Schramm G, Steglich W, Anke T, Oberwinker F (1978) Chem Ber 111:2779

3 Anke T, Schramm G, Schwalge B, Steffan B, Steglich W (1984) Liebigs Ann Chem: 1616

4 Grant DM, Paul G (1964) J Am Chem Soc 86:2984

5 Klebe G, BASF AG, Ludwigshafen, unpublished. We thank Dr. Klebe for the X-ray structure determination.

6 Anke T, Hecht HJ, Schramm G, Steglich W (1979) J Antibiot 32:1112

7 Schramm G (1980) Dissertation Universität Bonn

8 Meyer HH (1984) Liebigs Ann Chem: 791

9 Akita H, Koshiji H, Furuichi A, Horikoshi K, Oishi T (1983) Tetrahedron Lett 24: 2009

10 Musílek V, Cerná J, Sasek V, Semerdzieva M, Vondrácek (1969) Folia Mikrobiol (Prague) 14:377

11 Vondráček M, Čapková J, Šlechta J, Benda A, Musílek V, Cudlin J (1971) Czech Pat 136495; CA 75:4029n

12 Sedmera P, Musílek V, Nerud F, Vondráček M (1981) J Antibiot 34:1069

13 v Jagow G, Gribble GW, Trumpower B (1986) Biochemistry 25:775

14 Bäuerle J (1981) Dissertation Universität Tübingen

15 Anke T, Besl H, Mocek U, Steglich W (1983) J Antibiot 36:661

11 Vondráček M, Čapková J, Šlechta J, Benda A, Musílek V, Cudlin J (1971) Czech Pat 136495; CA 75:4029n
12 Sedmera P, Musílek V, Nerud F, Vondráček M (1981) J Antibiot 34:1069
13 v Jagow G, Gribble GW, Trumpower B (1986) Biochemistry 25:775
14 Bäuerle J (1981) Dissertation Universität Tübingen
15 Anke T, Besl H, Mocek U, Steglich W (1983) J Antibiot 36:661
16 Vondráček M, Vondračková J, Sedmera P, Musílek V (1983) Coll Czech Chem Commun 48:1508
17 Anke T, Scherer A, Steglich W, in preparation
18 Backens S, Steglich W, Bäuerle J, Anke T (1988) Liebigs Ann Chem:405
19 Nerud F, Sedmera P, Zouchová Z, Musílek V, Vondráček M (1982) Coll Czech Chem Commun 47:1020
20 Beautement K, Clough JM (1987) Tetrahedron Lett 28:475
21 Nakata T, Kuwabara T, Tani Y, Oishi T (1982) Tetrahedron Lett 23:1015
22 Kowalski CJ, Haque MS, Fields KW (1985) J Am Chem Soc 107:1429
23 Wittman MD, Kallmerten J (1987) J Org Chem 52:4303
24 Burke SD, Fobare WF, Pacofsky GJ (1983) J Org Chem 48:5221; Kallmerten J, Gould TJ (1983) Tetrahedron Lett 24:5177
25 Buss AD, Mason R, Warren S (1983) Tetrahedron Lett 24:5293
26 Akita H, Matsukara H, Oishi T (1986) Tetrahedron Lett 27:5397
27 Subík J, Behún M, Musílek V (1974) Biochem Biophys Res Commun 57:17
28 Douros J, National Cancer Institute, USA, personal communication.
29 Becker WF, v Jagow G, Anke T, Steglich W (1981) FEBS Lett 132:329
30 v Jagow G, Engel WD (1981) FEBS Lett 136:19
31 Brandt U, Schägger H, v Jagow G (1988) Eur J Biochem 173:499
32 Anke T, Schramm G, Steglich W, v Jagow G (1988) In: Kleinkauf H, v Döhren H, Jaenicke L (eds) The Roots of Modern Biochemistry. de Gruyter, Berlin New York
33 Schwalge B (1986) Dissertation, Universität Bonn; Schwalge B, Steglich W, Anke T: manuscript in preparation
34 Bushell MJ, Beautement K, Clough JM, Defraine P, Anthony VM, Godfrey CRA (1986) EP 178826 (ICI), CA 105:78670z; Schirmer U, Karbach S, Pommer EH, Ammermann E, Steglich W, Schwalge B, Anke T (1986) DE 3519282 (BASF AG), CA 106:80390x; Schirmer U, Karbach S, Pommer EH, Ammermann E, Steglich W, Schwalge B, Anke T (1986) DE 3519280 (BASF AG), CA 106:101890c
35 Weber W, Anke T, Scherer A, Steffan B, Steglich W: in preparation

LIPOPEPTIDES, AN INTERESTING CLASS OF MICROBIAL SECONDARY METABOLITES

J. Vater

Institut für Biochemie und Molekulare Biologie, Technische Universität Berlin, Franklinstr. 29, D-1000 Berlin 10

ABSTRACT

Lipopeptides represent class of microbial secondary metabolites which attained scientific, therapeutical and biotechnological interest. A short survey is given on the structures and activities of some representative lipopeptides. The present knowledge on the biosynthesis of such compounds is reviewed. The biosynthesis of surfactin from B. subtilis ATCC 21332 was studied. An enzyme was isolated and highly purified from a cell-free extract of this organism which catalyzes ATP-P_i-exchange reactions mediated by the amino acid components of this lipopeptide. The observed activation pattern is consistent with a peptide synthesizing multienzyme which activates its substrate amino acids simultaneously as reactive aminoacyl phosphates.

LIPOPEPTIDES - A SHORT SURVEY

Lipopeptides attain increasing scientific, therapeutical and biotechnological interest. The spectrum of their manifold activities covers antibiotics, antiviral and antitumor agents, immunomodulators as well as specific toxins and enzyme inhibitors. Many of these compounds also represent potent extracellular microbial surfactants. They are found widespread in the whole spectrum of microorganisms. At least 30 of such agents have been characterized in detail as far as their structure and activities are concerned. Many of these compounds are antibiotics showing a variety of antimicrobial properties. Their mechanism of action frequently is unknown. It seems obvious that their surface- and membrane-active character plays an important role in the expression of their specific activities. Therefore, the analysis of their physicochemical parameters and molecular structures as well as the identification of their targets in the

U. P. Schlunegger (Ed.)
Biologically Active Molecules

living cell are prerequisites for the elucidation of their mode of action.

Lipopeptides usually appear as mixtures of closely related variants due to differences in their lipid and/or peptide moieties. The variation in their fatty acid parts are not determined genetically, but rather depend on external factors, as nutritional conditions, choice of precursors or physicochemical parameters.

A collection of 10 representative lipopeptides is shown in Table 1. Their structures are demonstrated in Table 2. The majority of the lipopeptides so far known are produced by certain strains of streptomyces and bacilli. However, such compounds have also been isolated from fungi and cyanobacteria. Prominent lipopeptides from streptomyces species are the amphomycins (I) [1], enduracidins (II) [2-5] and globomycin (III) [6-8]. They are antibiotics with antibacterial activities which function as inhibitors of cell wall synthesis.

Among the lipopeptides demonstrated in Tables 1 and 2 the cyclosporins (IV) [9,10] have attained the highest therapeutical importance because of their immunosuppressive and anti-inflammatory properties [11]. Such agents are now used successfully as immunomodulators in organ transplantations, for example.

A few structurally related lipotetrapeptides with interesting properties have been isolated from the culture filtrates of some fungal strains. Chlamydocin (V) from Diheterospora chlamydosporia [12,13,14] cyclo (L-AOE→L-Aib→L-Phe→D-Pro), and a similar compound named WF-3161 from Petriella guttulata [23] cyclo (L-AOE→D-Phe→L-Leu→L-Pip), show efficient cytostatic properties and represent antitumor agents. The HC-toxin (VI) from Cochliobolus carbonum [15-17] cyclo (L-AOE→L-Pro→L-Ala→D-Ala) and Cyl-2 from Cylindrocladium scoparium [24,25] cyclo (L-AOE→D-O-MeTyr→L-Ile→L-Pip) are potent phytotoxins. HC-toxin is necessary for the exceptional virulence of the plant pathogen Cochliobolus carbonum race 1 on certain varieties of maize. All these lipotetrapeptides contain the unusual fatty acid 2-amino-9, 10-epoxy-8-oxodecanoic acid (AOE). Obviously the same structural motif can be used from four unrelated filamentous fungi.

Polymyxins (VII) [18] from Bacillus polymyxa are well-known antibiotics. Some of them are used therapeutically. Viscosin (VIII) [19] from Pseudomonas viscosa is an antiviral agent. Herbicolin A (IX) [20], an antibiotic from Erwinia herbicola which is highly active against yeasts and filamentous fungi, contains a D-glucose residue

linked in a 1-α-glycosidic bond to the fatty acid portion which is (R)-3-hydroxytetradecanoic acid. Herbicolin A constitutes the first glycolipodepsipeptide antibiotic so far known.
Also in blue-green algae (cyanobacteria) lipopeptides have been detected. The first compound of this kind, the structure and activity of which has been characterized in detail is cyanoginosin (X) [21, 22]. It was isolated from toxic water blooms of *Microcystis aeruginosa* which have been observed in several parts of the world. Cyanoginosin is a potent, rapidly acting hepatotoxin with acute toxicity to mammals. It is a cyclic heptalipopeptide which contains 3-amino-9-methoxy-10-phenyl-2,6,8-trimethyl-deca-4,6-dienoic acid (Adda) as exotic fatty acid component. Our research is primarily focussed on the biosynthesis of such lipopeptides and the analysis of the genes which code for their formation in the producer organisms. We are also engaged with the structural characterization and biotechnological usage of such compounds.

ENZYMATIC BIOSYNTHESIS OF BIOACTIVE PEPTIDES

Many bioactive peptides of microbial origin are formed nonribosomally [26-31]. Two enzymatic principles have so far been characterized for the biosynthesis of such compounds. Representative examples of the first pathway are the peptide antibiotics produced by *B. brevis* (the gramicidins and tyrocidines). They are formed on multifunctional polypeptide chains which activate their substrate amino acid in a two-step mechanism involving aminoacyl adenylates and thioesters as reactive intermediates. Multifunctional peptide synthetases of this type are equipped with a 4'-phosphopantetheine swinging arm as an internal transport system. The elongation of the growing peptide chain proceeds in a series of transpeptidation and transthiolation steps by interaction of the central SH-group of the carrier with the peripheral thiols at the reaction centers of the multienzyme (*thio-template-mechanism*). The intermediate peptides are covalently attached to the producer enzymes. They are vectorially transported within the enzyme structure. The sequence of such peptides is coded by the structural organization of the activation domains on the multifunctional polypeptide chain.
Representatives for the second reaction principle are the biosyntheses of glutathione [32],4'-phosphopantetheine [33] and mycobacillin [34,35], a tridecapeptide from *B. subtilis* B3. Here a one-step phos-

phate activation mechanism is used. In these processes an elongation of the growing peptide chain proceeds by sequential addition of non-activated acceptor amino acids to carboxyl activated intermediate peptides noncovalently attached to the reaction centers presumably as aminoacyl phosphates.

Only rather limited information is yet available for the biosynthesis of lipopeptides. Studies of this kind have been performed for polymyxin [36], cyclosporin A [37,38] and the HC-toxin [39]. The complete in vitro-synthesis has been obtained for polymyxin and cyclosporin A. In both processes a two-step activation (adenylation and thiolation) of the substrate amino acids has been observed indicating that the formation of these compounds is determined by the thiotemplate-principle. Both antibiotics are formed on large multifunctional enzymes. Polymyxin synthetase has a molecular mass of 640 kDa [36]. The biosynthesis of this peptide is initiated by the transfer of the octanoyl residue to this multienzyme by a separate acyl transferase and condensation with diaminobutyric acid. The molecular mass of the cyclosporin synthetase is approx. 700 kDa [37]. Here N-methylation of thio-esterified valine, leucine and glycine residues presumably is catalyzed by specific methyl transferases using S-adenosyl-methionine as substrate.

The HC-toxin seems to be controlled by a single gene locus. Walton [39] detected D-Ala and L-Pro dependent ATP-PP_i exchange activities which were catalyzed by two separable enzymes each with a molecular mass of approx. 310 kDa. Also for the biosynthesis of this phytotoxin presumably the thiotemplate-mechanism is used.

LIPOPEPTIDES PRODUCED BY VARIOUS STRAINS OF *BACILLUS SUBTILIS*

Our own work is concentrated on the biosynthesis of lipopeptides produced by various *B. subtilis* strains. Most of these compounds show antifungal activities. So far the structures of surfactin [40, 41] and six members of the iturin group (iturin A and C, bacillomycin D, F and L, mycosubtilin) have been characterized in detail by several authors [42-45]. The sequence of these agents which consist of seven amino acids in combination with a ß-hydroxy- or ß-amino-fatty acid is shown in Table 3. They appear as mixtures of closely related variants due to slight differences in their lipid portions [43,45]. The antibiotics of the iturin group show a characteristic LDDLLDL structure with a common part of the cyclic peptide: ß-amino-fatty acid→L-Asx→D-Tyr→D-Asn and a variable moiety containing four amino acid residues [45].

Surfactin is one of the most efficient biosurfactants so far known [46]. It shows antitumor activity against Ehrlich ascites carcinoma cells [47,48], inhibits fibrin clot formation [46] and functions as a potent inhibitor of cyclic adenosine 3',5'-monophosphate phosphodiesterase [49-51].

BIOSYNTHESIS OF SURFACTIN IN *B. SUBTILIS* ATCC 21332

We investigated the biosynthesis of surfactin in *B. subtilis* ATCC 21332. This lipopeptide was isolated from the culture filtrate by precipitation with HCl and purified by a combination of LH20-gel filtration, preparative thin layer chromatography and reversed phase HPLC. It was analyzed by FABMS and amino acid analysis. In vivo-incorporation of ^{14}C-labeled precursors into surfactin was controlled by hydrolysis (6 N HCl at 110°C for 24 h) [52]. The hydrolysates were analyzed by thin layer chromatography and radioscanning. It was demonstrated that all amino acid components were directly incorporated into the lipopeptide product. [^{14}C] acetate appeared in the fatty acid portion of surfactin. The main component is 3-hydroxy-13-methyltetradecanoic acid. In vitro biosynthesis of this lipopeptide was studied in *B. subtilis* ATCC 21322 which was fermented in the Landy medium [53]. Cells were harvested in the late exponential phase of the growth. In this period optimal rates of surfactin formation have been observed [54]. A cell-free system was prepared, as described previously [52]. In the crude extract an ammonium sulfate fractionation was performed. In the 40-55% fraction an enzyme activity was detected which catalyzed ATP-P_i exchange reactions mediated specifically by the surfactin components (L-Leu, D-Leu, L-Val, L-Asp, L-Glu). In Fig. 1 it is demonstrated that after AcA34 gel filtration these activities appeared in the same elution volume. The active fractions were pooled and loaded onto a DE-52 anion exchange column. The proteins were eluted with a salt gradient from 0-0,5 M NaCl. Also in this case the ATP-P_i exchange activities appeared at the same position in the gradient in a conductivity range between 12-16 $m\Omega^{-1}$. The active fractions were pooled again and loaded onto a Pharmacia Mono Q HR 5/5 column. A highly purified enzyme preparation was finally obtained by high resolution anion exchange fast protein liquid chromatography (FPLC), as shown in Fig. 2. These results indicate a surfactin synthesizing system which activates its substrate amino acids simultaneously. Presumably aminoacyl phosphates are involved in this process. Such an activation pattern would be different from the reaction sequence shown for glutathione and mycobacillin synthetase [32,34,35] which form reactive phosphates sequentially with the start amino acid and the peptide intermediates assembled in the elongation process. We now investigate, how the ß-hydroxy fatty acid is attached to the peptide moiety of surfactin in order to obtain the complete in vitro-synthesis of this lipopeptide.

REFERENCES

1 Bodanszky M, Sigler GF, Bodanszky A (1973) J Am Chem Soc 95:2352
2 Higashide E, Hatano K, Shibata M, Nakazawa K (1968) J Antibiot 21:126
3 Asai M, Muroi M, Sugita N, Kawashima H, Mizuno K, Miyake A (1968) J Antibiot 21:138
4 Hori M, Iwazaki H, Horii S, Yoshida I, Hongo T (1973) Chem Pharm Bull 21:1175
5 Iwazaki H, Horii S, Asai M, Mizuno K, Ueyanagi J, Miyake A (1973) Chem Pharm Bull 21:1184
6 Inukai M, Enokita R, Torikata A, Nakahara M, Iwado S, Arai M (1978) J Antibiot 31:410
7 Inukai M, Nakajima M, Osawa M, Haneishi T, Arai M (1978) J Antibiot 31:421
8 Nakajima M, Inukai M, Haneishi T, Terahara A, Arai M (1978) J Antibiot 31:426
9 Dreyfuss M, Härri E, Hofmann H, Kobel H, Pache W, Tscherter H (1976) Eur J Appl Microbiol 3:125
10 Rüegger A, Kuhn M, Lichti H, Lossli H-R, Huguenin R, Quiquerez C, von Wartburg A (1976) Helv Chim Acta 59:1075
11 Wenger RM (1985) Angew Chem 97:88
12 Closse A, Huguenin R (1974) Helv Chim Acta 57:533
13 Kawai M, Jasensky RD, Rich DH (1983) J Am Chem Soc 105:4456
14 Stähelin H, Trippmacher A (1974) Eur J Cancer 10:801
15 Walton JD, Earle ED, Gibson BW (1982) Biochem Biophys Res Commun 107:785
16 Liesch JM, Sweeley CC, Staffeld GD, Anderson MS, Weber DJ, Scheffler RP (1982) Tetrahedron 38:45
17 Gross ML, McCrery D, Crow F, Tomer KB, Pope MR, Ciufetti LM, Knoche HW, Daly JM, Dunkle LD (1982) Tetrahedron Lett 23:5381
18 Storm RD, Rosenthal KS, Swanson PE (1977) Ann Rev Biochem 46:723
19 Hiramoto M, Okada K, Nagai S (1970) Tetrahedron Lett 13:1087
20 Aydin M, Lucht N, König WA, Lupp R, Jung G, Winkelmann G (1985) Liebigs Ann Chem 1985:2285
21 Botes DP, Tuinman AA, Wessels PL, Viljoen CC, Kruger H, Williams DH, Santikarn S, Smith RJ, Hammond SJ (1984) J Chem Soc Perkin Trans 1:2311
22 Botes DP (1986) In: Steyn PS, Vleggaar R (eds) Mycotoxins and Phycotoxins p. 161 Elsevier, Amsterdam
23 Umehara K, Nakahara K, Kiyoto S, Iwami M, Okamoto M, Tanaka H, Kohsaka M, Aoki H, Imanaka H (1983) J Antibiot 34:478

24 Hirota A, Suzuki A, Suzuki H, Tamura S (1973) Agric Biol Chem 37:643
25 Hirota A, Suzuki A, Aizawa K, Tamura S (1973) Agric Biol Chem 37: 955
26 Lipmann F (1973) Acc Chem Res 6:361
27 Laland SG, Zimmer T-L (1973) Essays Biochem 9:31
28 Katz E, Demain AL (1977) Bacteriol Rev 41:499
29 Kleinkauf H, Koischwitz H (1978) Progr Mol Subcell Biol 6:59
30 Kleinkauf H, von Döhren H (1981) Current Topics Microbiol Immunol 91:129
31 Kurahashi K (1981) in: Corcoran JW (ed) Antibiotics, Vol IV, p. 325, Springer, Berlin Heidelberg New York
32 Meister A, Tate S (1976) Ann Rev Biochem 45:559
33 Akibo Y (1970) Methods Enzymol 18:350
34 Ghosh SK, Majumder S, Mukhopadhyay NK, Bose SK (1985) 230:785
35 Mukhopadhyay NK, Majumder S, Ghosh SK, Bose SK (1986) Biochem J 235:639
36 Kurahashi K, Komura S, Akashi K, Nishio C (1982) In: Kleinkauf H, von Döhren H (eds) Peptide Antibiotics-biosynthesis and functions, p. 275, de Gruyter,Berlin New York
37 Zocher R, Nihira T, Paul E, Madry N, Peeters H, Kleinkauf H (1986) Biochemistry 25:550
38 Billich A, Zocher R (1987) J Biol Chem 262:17258
39 Walton JD (1987) Proc Natl Acad Sci USA 84:8444
40 Kakinuma A, Hori M, Isono M, Tamura G, Arima K (1969) Agr Biol Chem 33:971
41 Kakinuma A, Sugino H, Isono M, Tamura G, Arima K (1969) Agr Biol Chem 33:973
42 Peypoux F, Guinand M, Michel G, Delcambe L, Das BC, Lederer E (1978) Biochemistry 17:3992
43 Peypoux F, Pommier M-T, Das BC, Besson F, Delcambe L, Michel G (1984) J Antibiot 37:1600
44 Peypoux F, Marion D, Maget-Dana R, Ptak M, Das BC, Michel G (1985) Eur J Biochem 153:335
45 Peypoux F, Pommier M-T, Marion D, Ptak M, Das BC, Michel G (1986) J Antibiot 39:636
46 Arima K, Kakinuma A, Tamura G (1968) Biochem Biophys Res Commun 31:488
47 Kameda Y, Matsui K, Kato H, Yamada T, Sagai H (1972) Chem Pharm Bull 20:1551
48 Kameda Y, Ouhira S, Matsui K, Kanatomo S, Hase T, Atsusaka T (1974) Chem Pharm Bull 22:938
49 Hosono K, Suzuki H (1983) J Antibiot 36:667
50 Hosono K, Suzuki H (1983) J Antibiot 36:674
51 Hosono K, Suzuki H (1983) J Antibiot 36:679
52 Kluge B, Vater J, Salnikow J, Eckart K (1988) FEBS Lett 231:107
53 Landy M, Warren GH, Rosenman SB, Colio LG (1948) Proc Soc Exp Biol Medicine 67:539
54 Vater J (1986) Progr Colloid Polymer Sci 72:12

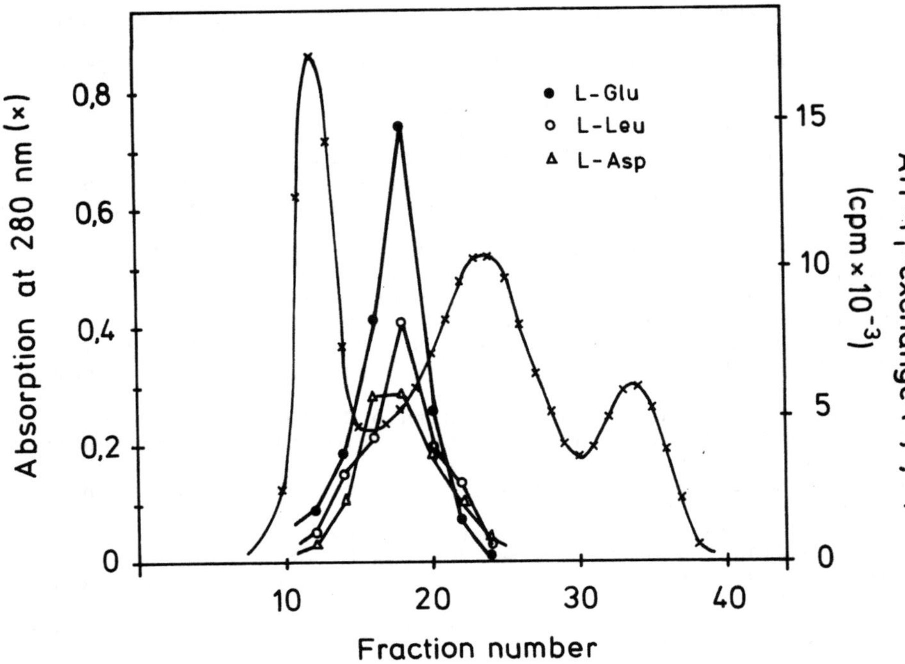

Fig. 1 AcA34 gel filtration of an enzyme which catalyzes ATP-P_i exchange reactions mediated by the amino acid components of surfactin. An ammonium sulfate fractionation was performed in a crude extract prepared from B. subtilis ATCC 21332. The fraction obtained in the range of 40-50% saturation of the precipitant was dissolved in a minimum volume of 50 mM Tris-HCl buffer, pH = 7,8; 0,5 mM EDTA and 1 mM dithioerythritol and loaded on the AcA34 column (53 x 2 cm). Elution was performed with the same buffer. Flow rate: 100 ml/h; temperature: 4°C.

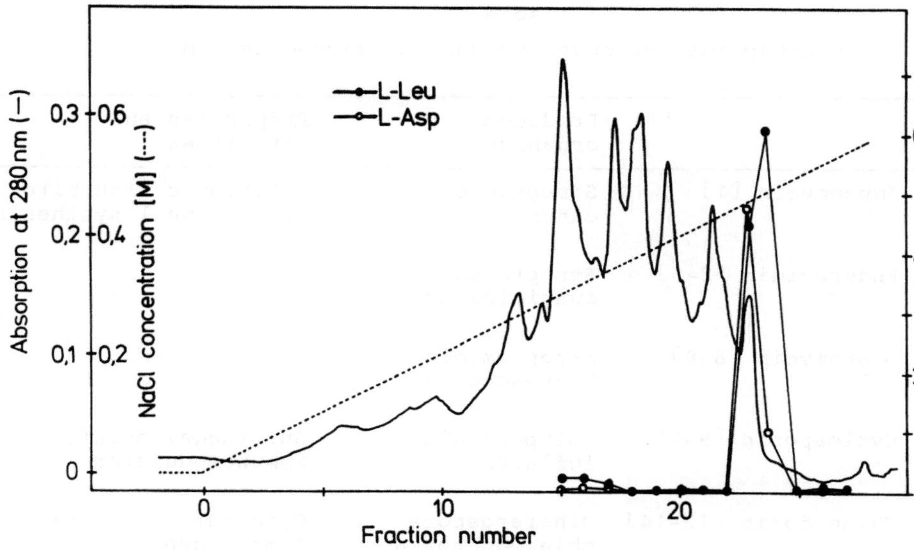

Fig. 2 Fast protein liquid chromatography (FPLC) on Pharmacia Mono Q HR 5/5 of an enzyme fraction activating the amino acid components of surfactin which was prepurified by DE-52 anion exchange chromatography. For elution of the proteins a gradient from 0-0,6 M NaCl in 50 mM Tris/HCl buffer, pH = 7,5; 2 mM dithioerythritol was used. Flow rate: 1 ml/min; temperature: 25°C.

Table 1

Lipopeptides produced by various microorganisms

Name		Producer organism	Properties and activities
I	Amphomycin [1]	Streptomyces canus	Antibiotic, inhibitor of cell wall synthesis
II	Enduracidin [2-5]	Streptomyces fungicidicus	"
III	Globomycin [6-8]	Streptomyces hygroscopicus	"
IV	Cyclosporin [9-11]	Tolypocladium inflatum	Antifungal agent, immunomodulator
V	Chlamydocin [12-14]	Diheterospora chlamydosporia	Cytostatic and anti-tumor agent
VI	HC-Toxin [15-17]	Cochliobolus carbonum race 1	Phytotoxin
VII	Polymyxin [18]	Bacillus polymyxa	Antibiotic
VIII	Viscosin [19]	Pseudomonas viscosa	Antiviral agent
IX	Herbicolin [20]	Erwinia herbicola	Antibiotic, highly active against yeasts and filamentous fungi
X	Cyanoginosin, Microcystin [21,22]	Microcystis aeruginosa (Cyanobacterium)	Hepatotoxin, acute toxicity to mammals

Table 2

Structures of lipopeptides compiled in Table 1

FA_1 = (+)-3-anteisotridecenoic or 3-isododecenoic acid; FA_2 = 10-methyl-undeca-2-(cis,4(trans)-dienoic acid; FA_3 = 2-methylamino-3-hydroxy-4-methyl-6-octenoic acid; FA_4 = 3-hydroxytetradecanoic acid; Adda = 3-amino-9-methoxy-2,6,8-trimethyl-10-phenyl-deca-4,6-dienoic acid; Dab = 2,4-diaminobutyric acid; Dab^e = D-erythro-2,3-diaminobutyric acid; Dab^t = L-threo-2,3-diaminobutyric acid; Pip = D-α-pipecolic acid; Abu = 2-aminobutyric acid; Sar = sarcosin; HyPhg; 2-amino-4-hydroxyphenylacetic acid or hydroxyphenylglycine; CHyPhg = 3-chloro- or 3,5-dichloro-HyPhg; Cit = citrulline; End = α(s)-amino-ß-4(R)-(2-imino imidazolidinyl)-propionic acid; aThr = allo-threonine; aIle = allo-isoleucine; Me Δ Ala = N-Methyl-dehydroalanine; ß-MeAsp = erythro-ß-methyl-D-iso-aspartic acid.

I Amphomycin

FA_1 → Asp → MeAsp → Asp → Gly → Asp → Gly → Dab^e → Val → Pro → Dab^t → Pip
(Pip linked back to Dab^t)

II Enduracidin A

FA_2 → Asp → Thr → D-HyPhg → D-Lys → D-aThr → HyPhg → D-HyPhg → aThr → Cit
HyPhg ← D-Ala ← End ← Gly ← D-CHyPhg ← D-Ser ← HyPhg ← D-End
(Cit → D-End; HyPhg → Thr)

III Globomycin

$CH_3(CH_2)_5CH(O)CH(CH_3)CO$ → MeLeu → aIle → Ser → aThr → Gly
(Gly linked to O)

IV Cyclosporin A

MeVal → FA_3 → Abu → Sar → MeLeu → Val
Val → MeLeu
MeVal ← MeLeu ← MeLeu ← D-Ala ← Ala ← MeLeu

VII Polymyxin E_1 (Colistin A)

6-CH_3-octanoyl → Dab → Thr → Dab → Dab → D-Leu → Leu
Thr ← Dab ← Dab ← Leu
(Thr → Dab)

VIII Viscosin

$CH_3(CH_2)_6CH(OH)CH_2CO$ → Leu → D-Glu → D-aThr → D-Val → Leu → D-Ser
D-Ser → Leu
D-aThr ← Ile ← D-Ser ← Leu

IX Herbicolin A

(R)-3-0(α-D-glucose)-FA_4 → Abu → L-Thr → D-aThr → D-Leu → Gly
Gly → D-Gln
L-Thr ← L-Arg ← L-Me-aThr ← Gly ← D-Gln

X Cyanoginosin

Adda → D-Glu → Me Δ Ala → D-Ala → Leu → ß-MeAsp → Ala
(Ala → Adda)

Table 3

Lipopeptides produced by various strains of <u>Bacillus subtilis</u>

Surfactin

C_{10-12}-CHCH$_2$CO→L-Glu→L-Leu→D-Leu→L-Val→L-Asp→L-Leu→D-Leu
O (ring closed between O and D-Leu)

Mycosubtilin

C_{13-14}-CHCH$_2$CO→L-Asn→D-Tyr→D-Asn→L-Gln→L-Pro→D-Ser→L-Asn
NH (ring closed between NH and L-Asn)

Iturin

C		L-Asp						
A	C_{11-13}-CHCH$_2$CO(NH) →	L-Asn	D-Tyr	D-Asn	L-Gln	L-Pro	D-Asn	L-Ser

Bacillomycin

D		L-Asn			L-Pro	L-Glu	D-Ser	
F		L-Asn			L-Gln	L-Pro	D-Asn	
L	C_{11-13}-CHCH$_2$CO(NH) →	L-Asp	D-Tyr	D-Asn	L-Ser	L-Gln	D-Ser	L-Thr

FAB: BASIC CONCEPTS AND PRACTICAL CONSIDERATIONS

Richard M. Caprioli

The Analytical Chemistry Center
and Department of Biochemistry and Molecular Biology
The University of Texas Medical School at Houston
Houston, Texas 77030

I. INTRODUCTION

The general theme of the conference, "from biological activity to structure", has prevailed in the biological research laboratory for decades. Although the methods used are now more sophisticated and the questions asked more detailed, the general approach to solving structural problems is much the same. Typically, biological activity is identified and the source and/or storage site of the material is located. Usually, an assay is developed to follow this activity through purification and isolation procedures. Chemical characterization is accomplished by a wide variety of techniques, including classical procedures as well as modern analytical methods in order to deduce the structure of the molecule responsible for this activity. It is often the case that from structure we return to biological activity in several ways. For example, this may be by chemical synthesis of the molecule in the laboratory in order to produce greater quantities of the compound at decreased cost, or by chemical modification of the original compound in order to increase its potency, to change its physical and chemical properties, or to decrease its side effects. It is through this cyclic process that we hope to identify chemical compounds that are important to life processes, learn how they interact with one another, and finally regulate and control these processes to our advantage. It is clear that this research approach is of great importance if we are to control disease and abnormal metabolic states. The appearance of deadly new diseases in the world makes this subject timely, and of vital concern to us all.

This presentation will focus on the characterization steps which help lead to the structural elucidation of biological molecules. In

U. P. Schlunegger (Ed.)
Biologically Active Molecules

a modern laboratory, many different types of methods and procedures are used to help uncover the structures of biomolecules. Most often isolation of the activity of interest is done first using methods which are based on extraction, chromatography, electrophoresis or other types of physical separation and purification procedures. The bioactive molecule can be followed by a bioassay and/or by a chemical or physical property of the compound. Over the years optical spectroscopy has played a key role in this regard, for example, an amino acid analysis for a protein involves the reaction of the amino group with ninhydrin to form a colored product, or the base analysis of nucleic acids which employs the absorption of light at a wave length of 254 nanometers. Detailed and specific structural characterization is accomplished by many different types of spectroscopic techniques and chemical reactions and degradations. Special procedures such as the Edman sequencing procedure for proteins, methylation analysis for carbohydrates and base sequencing procedures for nucleic acids are some examples. Modern instrumental methods commonly employed include mass spectrometry and nuclear magnetic resonance spectrometry. It is emphasized that all of these methods are used in combination to solve structural problems, and as the size and complexity of the compounds increase, so does the importance of these methodological interactions.

II. AN OVERVIEW OF MASS SPECTROMETRY

The use of mass spectrometry in the biological sciences goes back many years, beginning in the 1930's with stable isotope abundance measurements and the use of isotopes as tracers. In the 1950's, high resolution mass spectrometry allowed elemental compositions to be obtained and later in the 1960's, combined GC/MS allowed the biochemist to perform on-line separation and analysis of volatile compounds which were amenable to gas chromatography. In the late 1970's, LC/MS became of interest and different types of interfaces were developed, such as the moving belt and direct liquid injection techniques. In the 1980's, desorption ionization techniques have become widely used and these include fast atom bombardment, secondary ion, field desorption, thermospray and plasma desorption mass spectrometry, as well as a number of others still being developed. Most importantly, mass analyzer development has made significant advances with such innovations as laminated fast-scanning magnets, hybrid and tandem magnetic MS/MS systems, triple stage quadrupole MS/MS systems, and time-of-flight, ion cyclotron and ion trap devices.

Mass spectrometric data provides a number of different important parameters concerning the nature of a sample being analyzed. It can give the molecular weight and elemental composition of various compounds in the sample. Structure is often deduced from the mass spectrum, or at least, the presence of specific chemical groups, elements, or molecular type can be established. Quantitation can often be achieved either in a relative or absolute measurement through the use of an internal standard.

Desorption ionization techniques, also termed "soft" ionization techniques, have made an enormous impact on the biological research community. These include fast atom bombardment (1), secondary ion (2), Californium-252 plasma desorption (3), laser desorption (4), field desorption (5), desorption chemical ionization (6), thermospray (7), electrohydrodynamic (8) and electrospray ionization (9). Each has unique aspects and is currently under intensive research and development to further understand fundamental aspects and explore its utility. The most widely used, at least in the biological research community, is fast atom bombardment (FAB) mass spectrometry. Although the following discussion will concern itself with FAB, it is important to note that many basic considerations will also apply to a similar technique termed liquid secondary ion mass spectrometry (LSIMS).

III. FAST ATOM BOMBARDMENT MASS SPECTROMETRY

In FAB mass spectrometry, atomic particles are used to bombard the surface of a liquid sample. These particles are usually inert gases such as argon and xenon and typically have translational energies of 4-8 keV. Several types of atom guns have been developed, but perhaps the most common is the cold cathode discharge saddle field atom gun manufactured by Ion Tech Ltd. Basically, the production of ions in the gun takes place through electron impact processes in the presence of xenon gas, converting Xe atoms to Xe^+ ions plus an electron. These xenon ions are given kilovolts of energy and subsequently undergo neutralization of their charge by resonant charge exchange with xenon atoms and electron recombination processes. The gun is designed so that the atoms emerge from the nozzle of the device in a somewhat focused atomic beam containing about 10^{12} particles/sec. As these particles hit the liquid sample, molecules from the outer molecular layers of the sample are sputtered from the surface as shown in Figure 1 (10); those which are charged are transmitted into the analyzer of the mass spectrometer. In addition to individual molecules, molecular clusters are often sputtered and, indeed, these can be seen in the FAB mass spectrum of a sample. The mechanism of ionization is complex, but most certainly

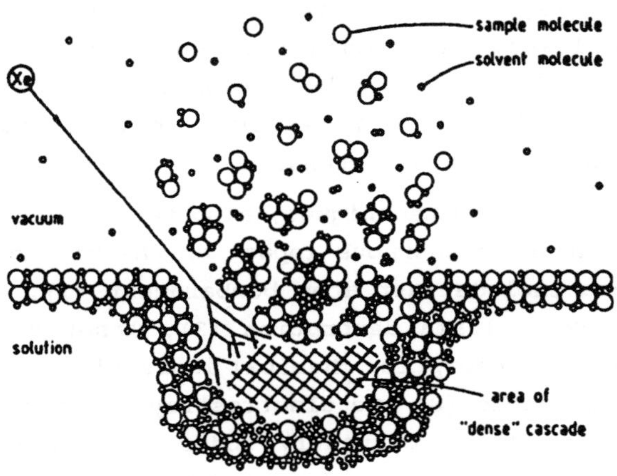

Figure 1. Schematic representation of the sputtering of surface layers of a liquid sample using FAB ionization. (Reprinted with permission from reference 10).

involves sputtering of preformed ions and ion molecule interactions in chemical ionization type processes. Thus, the overall analysis involves placing the liquid sample on the stage of the sample probe, inserting the probe into the mass spectrometer source through a vacuum lock, and subjecting the sample to atomic bombardment.

A. THE LIQUID SAMPLE MATRIX

Perhaps the most important characteristic of FAB is that it depends upon the sputtering of the surface of a liquid sample. The ability to deal with liquids opens up a wide variety of applications for the biochemist for the analysis of charged water-soluble compounds. As we will see later, this is a double-edged sword and gives rise to both good and bad aspects of the technique. High concentrations of a viscous organic liquid such as glycerol are used in order to minimize evaporation and prolong the liquid state of the sample in the high vacuum environment of the mass spectrometer. Typically, a sample is dissolved in 90 to 95% glycerol, thioglycerol, ethanolamine, or other suitable matrix. The advantages of the use of a liquid matrix in FAB are several. Perhaps first and foremost, it allows one to obtain mass spectra from underivatized compounds normally considered intractable for mass spectrometric analysis. In this regard, FAB is truly remarkable. In addition to maintaining the liquidity of the sample, use of a liquid matrix allows refreshment of the surface to take place as surface layers are sputtered during the analysis. In this way fresh sample from the interior of the liquid droplet is constantly being brought to the surface. Also, the liquid sample provides the opportunity to allow chemical reactions to be analyzed as they occur.

Some significant difficulties are encountered with the use of viscous organic matrices. They give rises to background chemical noise which appears as a peak at every mass in the mass spectrum. This is basically the result of radiation damage done to the organic components of the liquid matrix as a result of the bombardment process. Also, cluster ions are produced as background, that is, for example, glycerol gives rise to intense ions which are protonated clusters of glycerol. In adulterating the sample with an organic compound, one essentially 'contaminates' the sample and as a result the detection sensitivity for the analyte of interest is poor relative to other mass spectrometric ionization methods. The use of liquid solvents and matrices of different physical and chemical properties can give rise to distributions of molecules in a sample which may not be representative of the original sample. Surface interactions in the sample droplet are important and it has been shown that both hydrophilic and hydrophobic processes can give rise to concentration differences within the droplet. Finally, in a high vacuum environment, the sample concentration changes because of evaporation of the liquid. Nevertheless, despite the difficulties, FAB is an excellent ionization technique and is widely used for the mass spectrometric analysis of polar and charged molecules.

Many different liquid matrices have been used in FABMS, although some are targeted at samples which contain specific molecular types. Glycerol is commonly used in many cases with additives such as trifluoroacetic acid, salts, solvents and similar types of compounds to enhance overall ion yield and give good mass spectra. Thioglycerol is also widely used, particularly for high molecular weight molecules, i.e., for peptides having molecular weights above about 2,500 daltons. Other types of matrices include aminoglycerol (for glycopeptides), thioglycolic acid (for peptides), polyethyleneglycol (for saccharides), triethanolamine (for lipids), diethanolamine (for saccharides), 1,2,4-butanetriol (for phosphatides), and a dithiothreitol and dithioerythritol mixture in a 3 to 1 ratio (for organometallics) (11).

B. FAB MASS SPECTRA

A typical positive ion FAB spectrum of a peptide is shown in Figure 2. It is characterized by an intense molecular ion species at m/z 1182 which is an $(M+H)^+$ ion. Although cleavage of the molecule

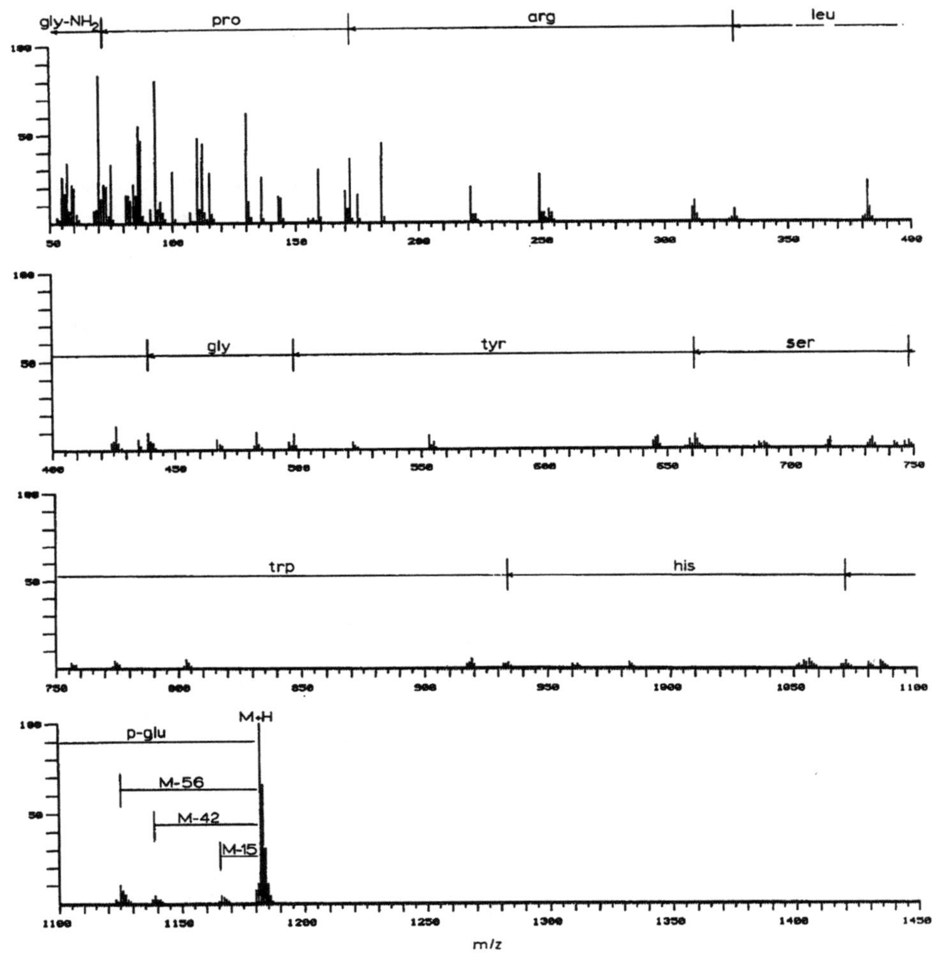

Figure 2. The positive ion FAB mass spectrum of a peptide fragment from the LHRH protein.

does occur, fragment ions are of relatively low intensity and, generally, sequence specific ions are not reliably observed in the FAB spectrum. Nevertheless, the production of an intense molecular ion from a completely underivitized and charged compound such as a peptide, in many cases up to molecular weights of 7,000 or 8,000 daltons, is extremely valuable to the biochemist. Molecular distributions of families of molecules are also seen using FABMS. Figure 3 shows the cationized molecular species of sphingomyelins from bovine erythrocytes (12). In this case, one can clearly see the distribution of the fatty acyl side chains; saturated C16, C18, C22 and C24 side chains and a C24 with one unsaturation.

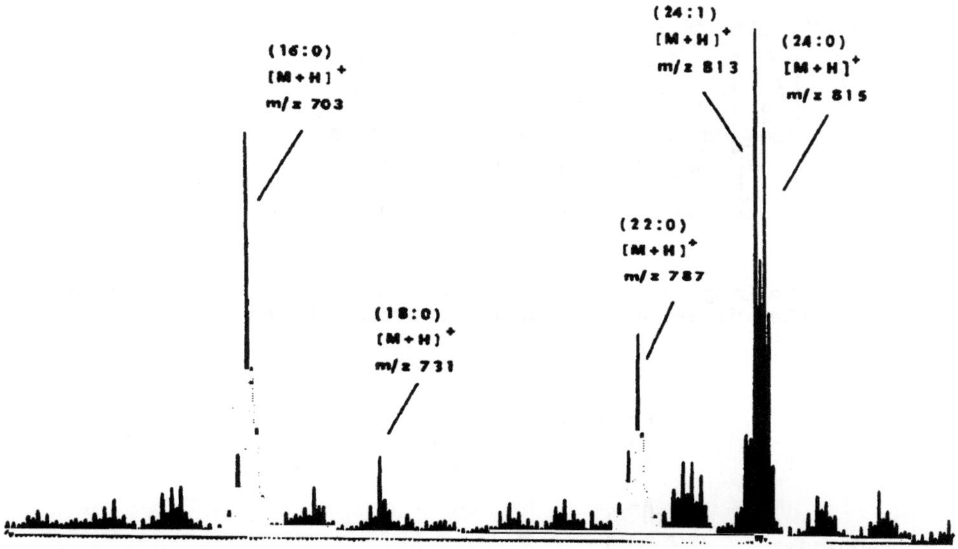

Figure 3. FAB spectrum of the cationized molecular species of sphingomyelins of bovine erythrocytes. (Reprinted with permission from reference 12).

For certain types of molecules, especially those carrying excess negative charges, negative ion FAB is preferred. Figure 4 displays such a spectrum for an oligonucleotide, showing fragment ions of relatively high intensity which can be used to help establish the sequence of the bases in the molecule (13). The choice of positive or negative ion acquisition can be crucial when analyzing some compounds. Figure 5 shows an example of two peptides which are analyzed by positive and negative ion FAB (14). The peptides are present in equal molar concentrations and, indeed, one records approximately a 1:1 ratio for $(M-H)^-$ in negative ion FAB. Positive ion FAB, however, shows only one molecular species. Examination of the sequence of these two peptides would lead to the prediction

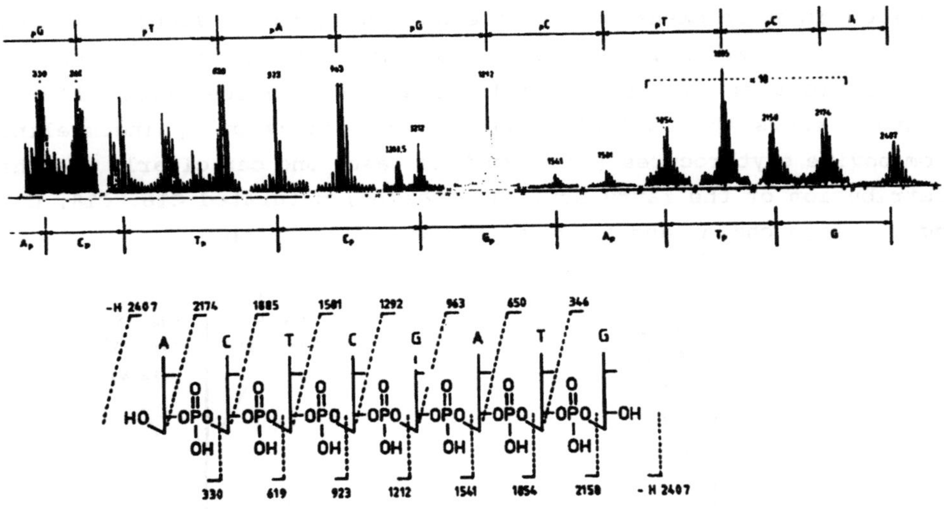

Figure 4. The negative ion FAB spectrum of an oligonucleotide. (Reprinted with permission from reference 13).

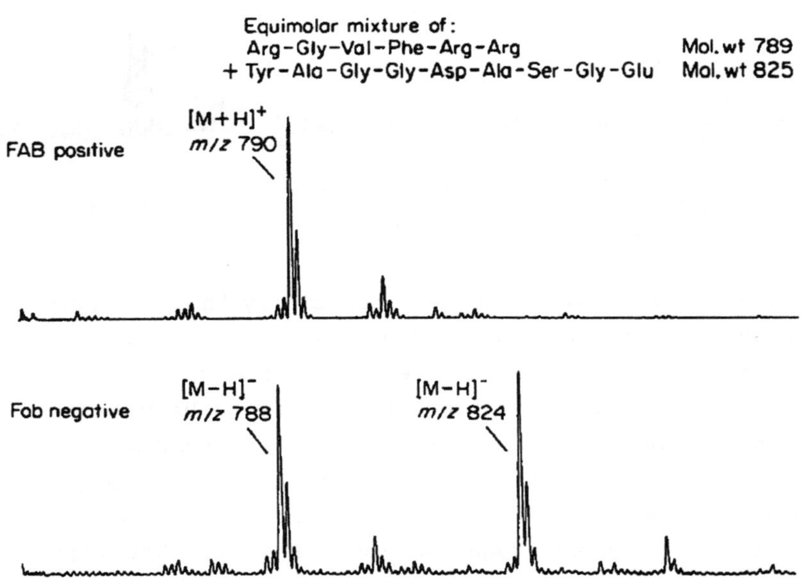

Figure 5. Positive and negative ion FAB of a mixture containing equal concentrations of a basic and an acidic peptide. (Reprinted with permission from reference 14).

of different charge states in solution and would suggest that they would both not form positive and negative ions to the same extent.

In the interpretation of FAB mass spectra, it is important to recognize the effects of the background and the practical limitations and performance of the technique. First, consider the background produced by glycerol, shown in Figure 6 (15). Glycerol cluster ions

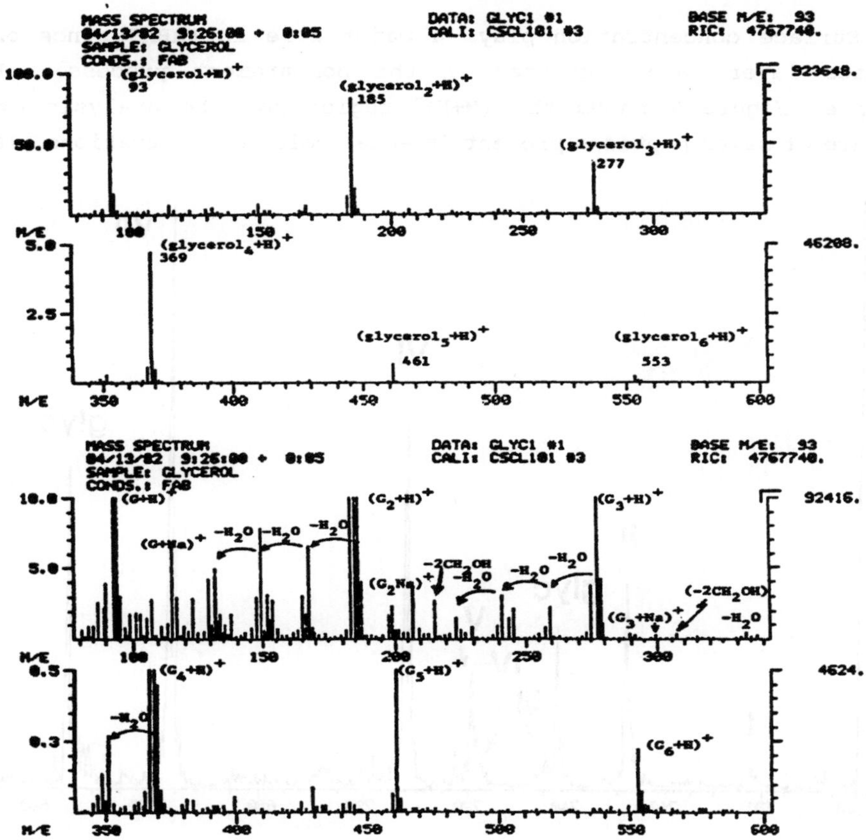

Figure 6. The positive ion FAB spectrum of glycerol. The lower panel is plotted at higher gain. (Reproduced with permission from reference 15).

are found at masses which correspond to multiples of the molecular weight of glycerol (92 daltons) plus as proton, i.e., ions are recorded at m/z 93, 185, 277, 369, 461, etc. In addition, the bottom portion of Figure 6 shows the same spectrum at higher sensitivity where many fragment ions are seen that are derived from these glycerol clusters. Loss of water is commonly observed both from the clusters and fragments, although loss of CH_3O fragments also occurs. The overall background from these cluster ions and radiation damage to glycerol severely limit the sensitivity of FAB, rendering it less sensitive than other mass spectrometric ionization techniques such as electron impact and chemical ionization.

C) SURFACE EFFECTS IN LIQUID SAMPLES

Surface concentration plays a major role in FABMS since only surface layers are sputtered in the bombardment process. For example, Figure 7 shows the $(M+H)^+$ region for the analysis of a mixture of seven peptides present in equal molar concentrations (16).

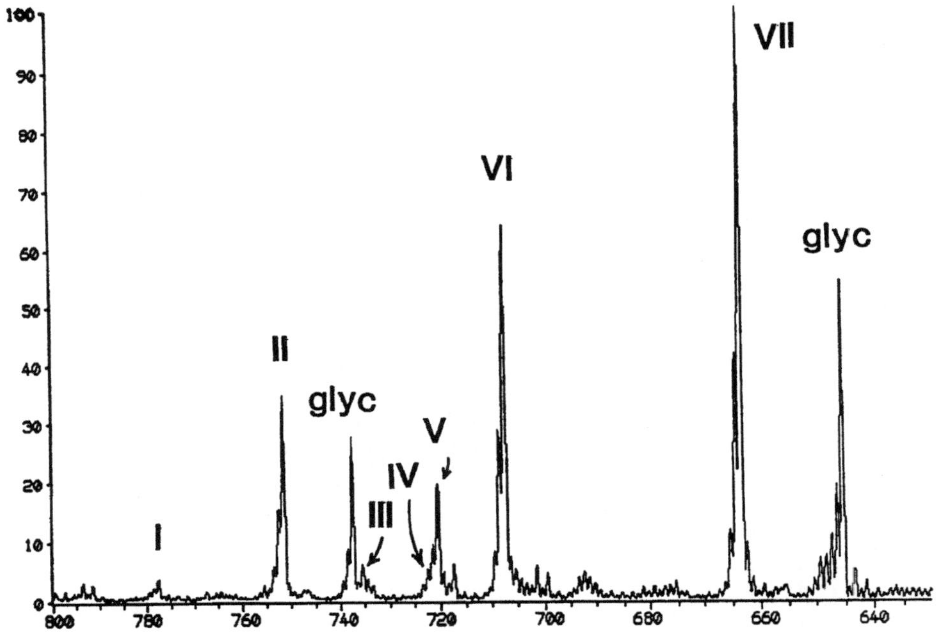

Figure 7. FAB spectrum of a mixture of seven heptapeptides which differ in sequence only in the central three residues. (Reproduced with permission from reference 16).

These peptides have different net charges in solution, as well as different hydrophobic/hydrophilic indices, contributing factors which do not allow them to occupy the surface in proportion to their concentration. These effects and others which can lead to discrimination of the species measured at the surface should be kept in mind in interpreting FAB mass spectra. Finally, although we have been discussing protonated molecular species [$(M+H)^+$ions], fast atom bombardment spectra are often characterized by other cationized addition products, for example, $(M+Na)^+$ ions in the presence of sodium salts. In simple mixtures, these adduct ions are easy to interpret and can aid in identification of certain species.

IV. MASS ANALYZERS

Although this presentation is primarily concerned with the fast atom bombardment ionization process, a few comments about mass analyzers and how they are used today in the analysis of biological compounds by mass spectrometry are in order. The two most commonly used analyzers are the magnetic and quadrupole devices. Generally, magnetic instruments are especially useful in applications requiring the spectrum of a compound whose molecular weight is over about 3,000 daltons or where high resolution is required. The practical mass limitation is about 10,000 daltons, although several exceptions have been described. For example, using a Cs^+ ion gun and a high sensitivity array detector, the $(M+H)^+$ ions for bovine trypsin (MW 23,290) has been recorded on a high field magnetic instrument (17). Quadrupole instruments are useful for lower mass range applications with electron impact and chemical ionization sources, and up to approximately 2,000 to 3,000 daltons with FAB. These instruments are relatively fast scanning devices and are extremely useful for GC/MS applications, particularly for high efficiency capillary column separations.

Recent advances in ionization techniques which give higher molecular weight species has led to a renewed interest in time-of-flight mass spectrometry principally because of its practically unlimited mass range. Using plasma desorption and laser desorption techniques, time-of-flight instruments have achieved remarkable results, recording ions of over 50,000 daltons from proteins and in one case, of approximately 200,000 daltons for an enzyme complex (18). However, they are relatively low resolution instruments. Other types of mass analyzers such as the ion cyclotron resonance and trap/ion storage devices, Wein filters and other analyzers are also under intensive study in attempts to further develop these techniques.

Over the past ten years, MS/MS instrumentation has been developed and has been shown to be particularly effective in the analysis of compounds derived from biological systems. Figure 8 shows the MS/MS configuration of the triple stage quadrupole instrument and a comparison to GC/MS (19). Basically, in a GC/MS analysis, one uses the gas chromatograph as a separation process with an on-line interface directly coupled to the mass spectrometer so that

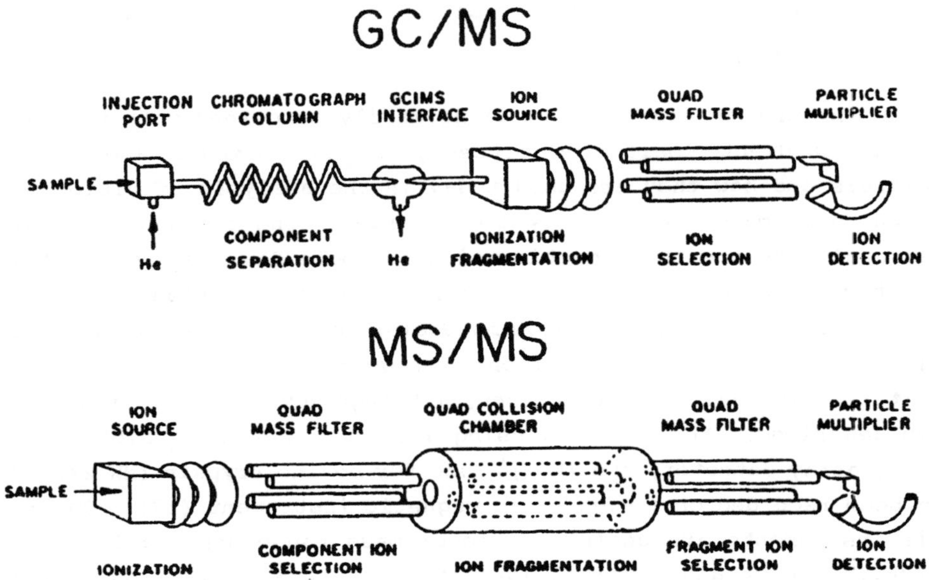

Figure 8. Comparison of the principles of operation of GC/MS and MS/MS. (Reprinted with permission from reference 19).

individual components which are eluted from the chromatograph are ionized and mass analyzed. The MS/MS combination conceptually works the same way. For a mixture analysis, the ion source produces $(M+H)^+$ ions from the molecular species present in the sample. The first mass filter is used to select one of these molecular species and transmit it into the collision chamber. Here, collision activated decomposition (CAD) fragments the molecule and in the third quadrupole, these fragments are individually analyzed to give a daughter ion spectrum. Thus, one can observe fragment ions from a specific molecular ion even though it is derived from a compound which was introduced into the source of the instrument as a mixture. Another type of MS/MS is the tandem high resolution system which basically works in the same manner and employs high resolution double-focusing mass spectrometers as the two mass analyzers. This type of instrument gives high energy collisions and has been shown to be effective in the sequencing of peptides.

V. CALCULATION AND MEASUREMENT OF MOLECULAR WEIGHTS BY FABMS

Several considerations come into play when measuring the molecular weight of compounds with a technique such as FABMS. For example, the molecular ion region of human insulin at approximately 5,780 daltons, when recorded at unit resolution, shows a series of molecular peaks across this region of the mass spectrum which are shifted several mass units higher than the $(M+H)^+$ calculated from the nominal molecular weight of insulin. Two factors are responsible for these effects.

First, elements are not integral masses but have fractional masses, that is, mass defects. Thus, although carbon is set at 12.000 amu by convention, hydrogen has a mass of 1.00782, nitrogen 14.0031, oxygen 15.9949, sulphur 31.9721, etc. For molecules of low molecular weight, these mass defects do not significantly alter the mass of the molecule and are generally ignored. At higher masses, mass defects can lead to appreciable mass increments. For example, if one considers the insulin molecule which consists of C_{256}, H_{382}, N_{65}, O_{76}, S_6, the exact monoisotopic molecular weight is calculated to be 5,773.45 ($[M+H]^+$ = 5,774.45), while that obtained from calculations using nominal elemental masses is 5,772.

The second factor which must be considered is the natural abundance of stable isotopes, i.e., the carbon-13 abundance is approximately 1.11%; nitrogen-15, 0.37%; and oxygen-18, 0.20%. At high molecular weights measured at unit resolution, the isotopic distribution of the molecular species can become quite broad. Thus, for a molecule containing C_{153}, H_{224}, N_{420}, O_{50}, S_1, the monoisotopic mass is 3,481.7 and the spectrum shows four significant peaks above this mass as seen in Figure 9. The average mass of the isotope cluster for $(M+H)^+$, however, is approximately 3,483.7. The $(M+H)^+$ ion distribution is further exaggerated at higher mass, for example, for a protein at approximately mass 5,780 where nine peaks are recorded, with an monoisotopic mass of 5,774.5 daltons and an average mass of approximately 5,778 daltons. In addition, each one of the mass peaks can be in itself extremely complex. In the last example, the composition of the most abundant peak in the molecular ion cluster, m/z 5,777 is composed of 13 individual isotopic molecules, although only four or five are of major intensity.

<u>*Natural Isotope Abundance*</u>

^{2}H	0.015 %
^{13}C	1.11
^{15}N	0.37
^{18}O	0.20

$C_{153}H_{224}N_{42}O_{50}S$

$C_{256}H_{382}N_{65}O_{76}S_{6}$

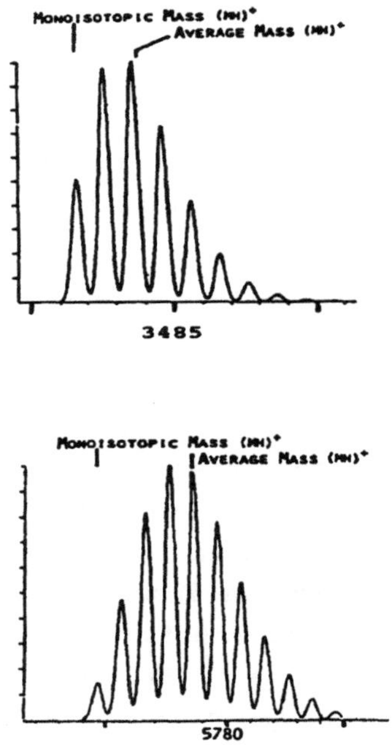

Figure 9. Isotopic distribution of the molecular species of two polypeptides as they would be recorded by FABMS at unit resolution.

For the most part, measurements of high molecular weight compounds (>3000 daltons) are made at less than unit resolution in order to maximize the signal intensity. Decreasing the resolution of a magnetic instrument is accomplished by opening up the source and collector slits, and as a consequence, more ions are allowed to pass and be recorded. In these measurements, such as that for insulin shown in Figure 10, the gaussian-shaped peak recorded represents the unresolved isotope cluster. A centroid fit of this peak was done to obtain the average mass of the $(M+H)^{+}$ ions and this was measured to be 5,778.96 daltons, in very good agreement with the calculated average mass for $(M+H)^{+}$ of 5,778.59. The nominal molecular weight calculates to be 5,772.

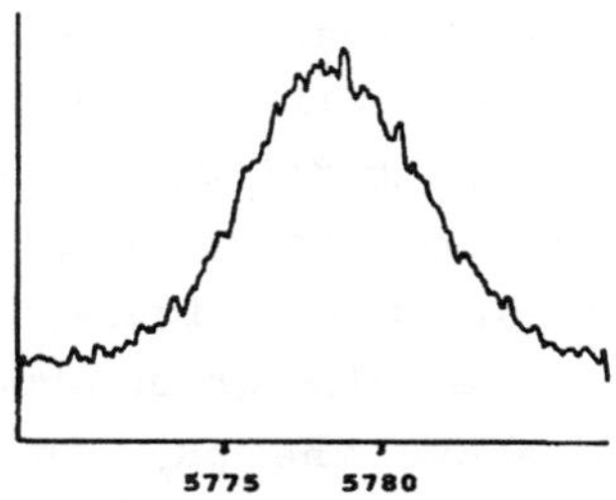

$C_{256}H_{382}N_{65}O_{76}S_6$	*nominal MW*	5772
	monoisotopic (M+H)$^+$	5774.45
	average (M+H)$^+$	5778.96

Figure 10. FAB mass spectrum of the molecular ion region of procine insulin at a resolution of about 1000. (Reprinted with permission from reference 20).

VI. SOLUTION CHEMISTRY AND FAB

One of the consequences of the liquid sample used to obtain a FAB spectrum is that it is possible to directly observe solution chemistry. Figure 11 shows a simple example of this for the effect

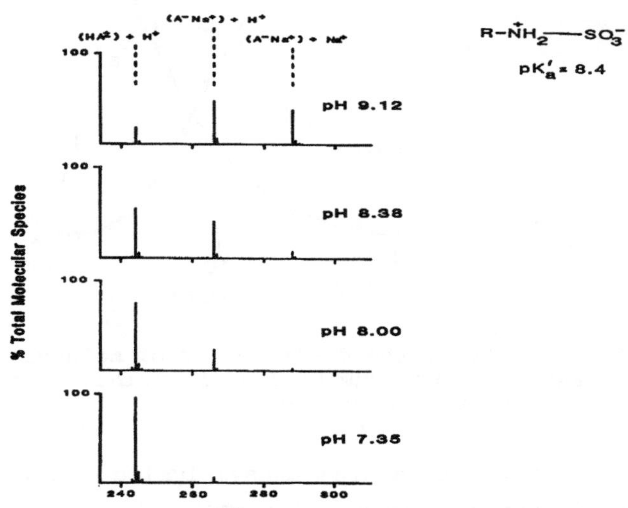

Figure 11. FAB mass spectrum of the molecular ion region of a zwitterionic organic salt recorded at four pH values. (Reprinted with permission from reference 21).

of changing the pH of a solution containing a zwitterionic compound with a pK_a of approximately 8.4 (21). The four panels show the molecular ion distribution at four different pH values, 9.12, 8.38, 8.0 and 7.35. From considerations of solution chemistry, one would expect the results shown in this Figure, i.e., at pH 7.35, significantly below the pK_a, the amine group should be protonated and at pH 9.12, significantly above the pK_a, the amine group should mainly be unprotonated. This observation is interesting because it shows that at least for some acids under rigorously controlled conditions, gas phase ions which are sputtered from liquid samples can accurately reflect the molecular distributions which existed in the liquid sample itself. This also is seen qualitatively when salt is present in a solution undergoing FAB analysis. Figure 12 shows the effect of increasing the concentration ratio of sodium to

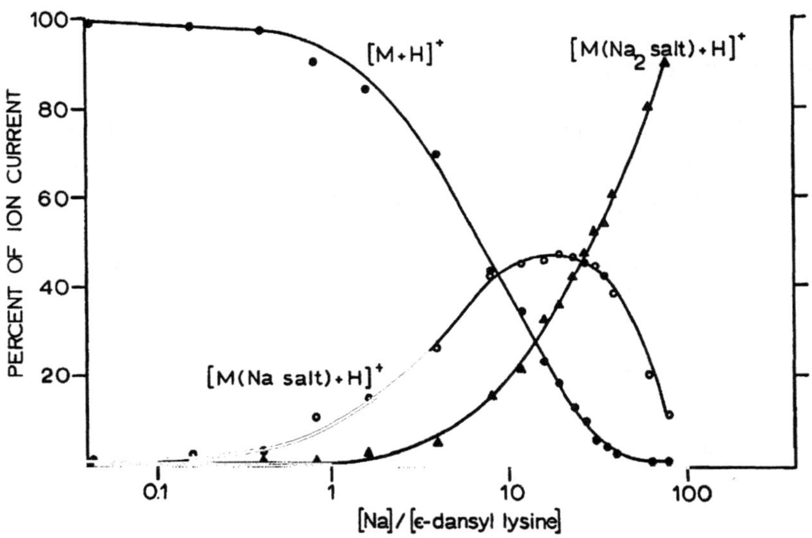

Figure 12. Effect of Na^+ on the distribution of molecular species in the FAB mass spectrum of dansyl lysine. (Reprinted with permission from reference 22).

analyte in a solution containing dansyl-lysine (22). As the Na^+ concentration increases, $(M+H)^+$ decreases in intensity with a concomitant increase in sputtered ions which contain the protonated sodium salt and the sodium cationized species. Thus, it appears that ions which are associated with one another in solution can be sputtered as ionic clusters. However, it should be pointed out that these effects appear to be complex and accurate ionic distributions may only be measured in this way under specific and well-controlled conditions.

VII. QUANTITATION

Quantitation using FAB has been demonstrated by a number of investigators under a variety of conditions. If one considers the quantitative aspects of an analysis done by mass spectrometry, several assumptions are normally made; 1) ion current is proportional to concentration, 2) ion current is unaffected by the presence of other compounds, and 3) chemically similar compounds have similar ionization efficiencies. With respect to FAB, the evaluation of these considerations is not straightforward. This is illustrated in Figure 13 by the stable isotope dilution analysis of γ-aminobutyric acid denoted as GABA-d_0 and its dideuteriated analog, GABA-d_2 (23).

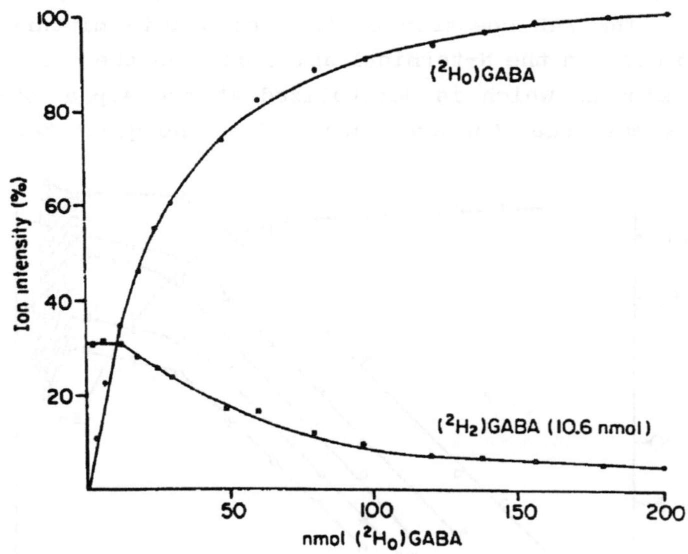

Figure 13. Effect of increasing concentrations of GABA-d_0 on ion intensity of GABA-d_0 and GABA-d_2. The d_2 species was present in all samples at a level of 12 nanomoles. (Reprinted with permission from reference 23).

In this analysis, GABA-d_0 was increased in concentration from approximately 2 to 200 nanomoles while keeping the GABA-d_2 concentration constant at 12 nanomoles in each sample. It is seen that in the initial portion of the curve, presumably when the surface of the sample droplet is not saturated with analyte, both species appear to behave independently. As the surface approaches saturation, the signal from the increasing concentration of d_0 species begins to level off and does not increase in proportion to its concentration. For the d_2 species, as the surface approaches saturation, the signal actually decreases. Thus, in terms of a normal standard response curve, this result appears atypical and is the consequence of sputtering from a fixed surface area. However, if one plots the ratio of the two ion intensities versus the ratio of their actual concentrations, a constant is obtained throughout the range and, in fact, gives an excellent quantitative correlation.

In a practical example, quantitative analysis using an internal standard is illustrated by the determination of the N-terminal amino acid of the P-22 phage tail protein using the dansylation technique (22). Briefly, the protein was first treated with dansyl chloride, hydrolyzed with acid to form amino acids, and the dansyl amino acids identified. The FAB spectrum of the amino acid mixture would be expected to contain the N-terminal amino acid as the only one of the dansylated species which is derivatized at the alpha amino group. Figure 14 shows the standard curves for the quantitation of the

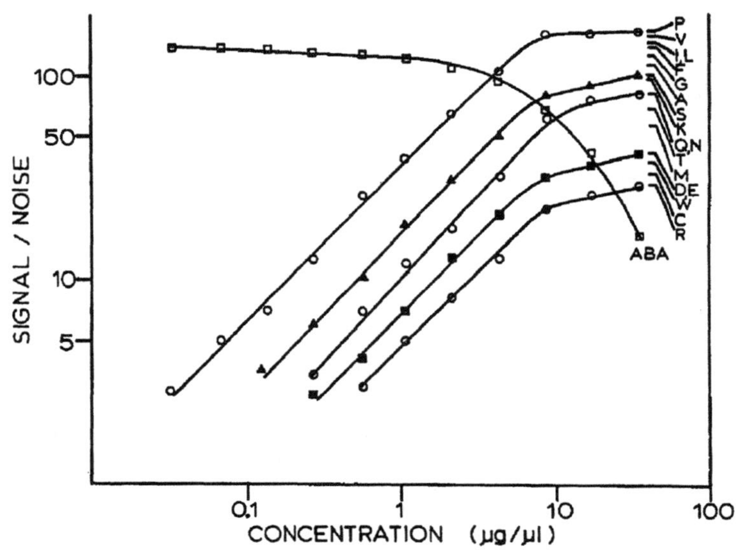

Figure 14. Standard response curves for the quantitation of dansyl-amino acids using dansyl-aminobutyric acid as an internal standard. (Reprinted with permission from reference 22).

various dansyl amino acids using dansyl aminobutyric acid as an internal standard. Once again, the curves show that with increasing concentration of dansyl amino acid, the signal saturates, but for the internal standard which is kept at a constant amount, the signal decreases. Data from the analysis of the dansylated phage P-22 tail protein showed several dansyl amino acids to be present. However, dansyl threonine was the only unexpected amino acid and thus was the N-terminal residue. (Mono-dansyl lysine and mono-dansyl tyrosine are expected from reactions of the side chains of these residues in internal positions in the chain.) Calculation of the amount of danysl threonine from the ratio of $(M+H)^+$ for this compound and the internal standard showed that greater than 96% of the dansyl threonine was recovered and indicated that the protein was relatively pure and contained a single amino acid in the N-terminal position.

VIII. SUMMARY

The development and use of soft ionization techniques, particularly FABMS, has proved to be of significant value in the analysis of molecules of biological origin. The use of the liquid sample is critical in the performance of this technique and provides the opportunity to use the technique to directly analyze ongoing biochemical reactions. The ability to analyze relatively large biomolecules in their native aqueous state has brought mass spectrometry into a new level of utility for biological investigators. In essence, it opens up the realm of mass specific identification and detection to a wide range of molecules including some medium molecular weight proteins, enzymes and other compounds derived from biological systems.

References

1 Barber M, Bordoli R, Elliott GJ, Sedgwick R, Tyler A (1982 Anal Chem 54:645A
2 Day RJ, Unger SE, Cooks RG (1980) Anal Chem 52:557A
3 Macfarlane RD (1983) Anal Chem 55:1247A
4 Kistmaker PG, vander Pyle GJ, Haverkamp J (1981) In: Morris HR (ed) Soft Ionization Biological Mass Spectrometry. Heyden, London 120
5 Wood GW (1982) Mass Spectrom Rev 1:63
6 Cotter RJ (1980) Anal Chem 52:1589A
7 Vestal ML (1983) Mass Spectrom Rev 4:447
8 Cook KD (1986) Mass Spectrom Rev 4:467
9 Dole M, Mack LL, Hines RL, Mobley RC, Ferguson LD, Alice MD (1968) J Chem Phys 49:2240
10 Wong SS, Rollgen FW (1986) Nucl Inst and Methods in Phys Res B14:436
11 Grower JL (1985) Biomed Mass Spectrom 12:191
12 Lehmann WD, Kessler M, (1983) Chem Phys Lipids 32:123

13 Grotjahn L, Frank R, Blocker H (1982) Nucl Acids Res 10:4671
14 Lehmann WD, Kessler M, Konig WA (1984) Biomed Mass Spectrom 11:217
15 Caprioli RM, Beckner CF, Smith LA (1983) Biomed Mass Spectrom 10:94
16 Caprioli RM, Moore WT, Petrie G, Wilson K (1988) Int'l J Mass Spectrom Ion Phys (in press)
17 Ashcroft AE, Buchanan RA, Elliott GJ, Evans S, Milton DJ, Wright B, Walls FC (1988) Proceedings of the 36th ASMS Conf on Mass Spectrom, June 5-10, San Francisco 1156
18 Hillenkamp F (1988) presented at the 11th Int'l Conference on Mass Spectrometry, Bordeaux, Aug 29-Sept 2
19 Johnson JV, Lee MS, Lee MR, Brotherton HO, Yost RA (1986) In: Gaskell SJ (ed) Mass Spectrometry in Biomed Res. Wiley, New York, 459
20 Fenselau C, Yergey J, Heller D (1983) Int'l J Mass Spectrom Ion Phys 53:5
21 Caprioli RM (1983) Anal Chem 55:2387
22 Beckner CF, Caprioli RM (1983) Anal Bchm 130:328
23 Beckner CF, Caprioli RM (1984) Biomed Mass Spectrom 11:60

CONTINUOUS-FLOW FAST ATOM BOMBARDMENT MASS SPECTROMETRY

Richard M. Caprioli

The Analytical Chemistry Center
and Department of Biochemistry and Molecular Biology
The University of Texas Medical School at Houston
Houston, Texas 77030

I. INTRODUCTION

Fast atom bombardment (FAB) mass spectrometry has made an enormous impact on analytical biochemistry, opening up new approaches to the biochemist and molecular biologist for the solution of structural problems. Although FAB has become widely used and has many advantages over other mass spectrometric ionization techniques, it is clear that several disadvantages of the technique severely limit its utility. Generally, these difficulties are a consequence of the high concentration of glycerol or other organic liquid support matrix used in the technique.

Continuous-flow FAB (CF-FAB) was developed as a means of interfacing liquid chemistry to mass spectrometry and minimize difficulties encountered with the use of organic matrices (1, 2). Perhaps the greatest asset of CF-FAB is that it is compatible with completely aqueous samples and allows the biochemist to directly analyze isolated samples and reaction solutions. Low concentrations of organic liquid matrix are used and are included in the carrier solution. CF-FAB provides significant improvements in performance in regard to parameters such as sensitivity, ion suppression effects, background chemical noise and matrix cluster ion formation. These improvements will be described in detail later.

II. PERFORMANCE

A. OPERATIONAL CONSIDERATIONS

Figure 1 shows a CF-FAB probe interface which was developed for a magnetic high resolution mass spectrometer. It consists of a sample introduction probe through which a narrow bore fused-silica

U. P. Schlunegger (Ed.)
Biologically Active Molecules

capillary is passed; one end of the capillary terminates at the probe tip where the bombardment process occurs and the other end at a micro-injector valve through which samples can be introduced into a flow of carrier solution supplied by a pump. The micro-injector valve may be bypassed for continuous monitoring applications. The polyimide coated fused-silica capillary is necessary when such a probe is used in magnetic mass spectrometers because the sample stage (probe tip) is held at high voltage.

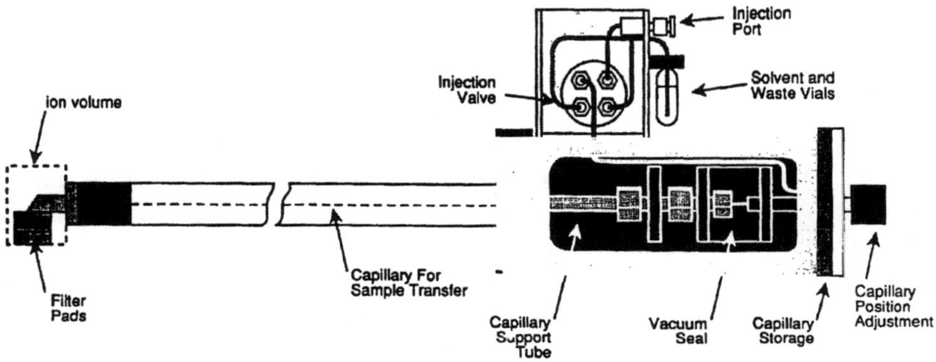

Figure 1. Continuous-flow FAB probe designed for a magnetic high resolution mass spectrometer. (Courtesy Finnigan MAT Corp.)

The basic principal of operation of the CF-FAB probe involves the delivery of aqueous sample solution to the sample stage and subsequent removal of the liquid at approximately the same rate, leaving only a thin fluid surface for bombardment. Formation of gaseous ions occurs through bombardment of this thin liquid film with energetic xenon atoms. Removal of the liquid from the surface can be accomplished by evaporation through gentle heating and by drawing the liquid into a filter paper pad or other reservoir by capillary action. Evaporation of the liquid involves the balance of the rate of flow of the liquid and also the temperature of the probe tip to obtain optimal operating conditions. Standard pumping capacities of most commercial mass spectrometers are sufficient to handle evaporation of up to about 10 μL/min of water, acetonitrile or other volatile solvent. Using the wicking action of filter paper, removal of liquid from the probe tip has been achieved at flow rates of up to approximately 15-20 μL/min. Typically we have found that excellent performance is obtained using the CF-FAB probe when a flow rate of 5-10 μL/min is used, allowing the device to operate for 6-8 hours or more without being removed from the spectrometer. Figure 2 shows a typical injection profile of approximately 100 pmol of the peptide substance P obtained with CF-FAB analysis using a Kratos MS50 mass spectrometer. The left-hand portion of the Figure shows the

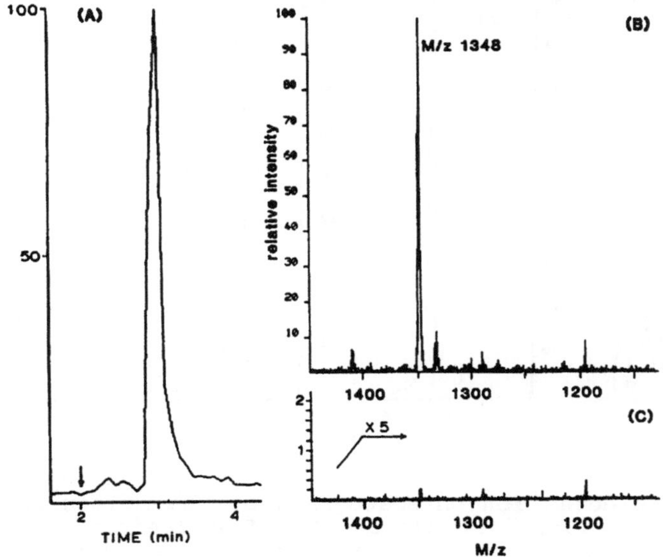

Figure 2. CF-FAB analysis of an injection (at arrow) of 13.5 ng (10 pmol) of the peptide Substance P. A) Total ion chromatogram; B) MS at peak maximum; C) MS 1 min after peak maximum. (Reproduced with permission from reference 1).

total ion chromatogram of the peak injection and the right-hand portion the mass spectra obtained. The top mass spectrum was taken from the chromatographic peak maximum located at approximately three minutes on the chromatogram time scale (sample was injected at 2 min) and the bottom mass spectrum in the figure taken at approximately four minutes. It is seen that at the four minute time point, the memory or carry-over of peptide signal at m/z 1348 is negligible, that is, less than 0.05% of the ion current obtained from the peak maximum remains one minute after the peak maximum.

The reproducibility of peak injections, using the same sample, can vary depending upon the stability conditions achieved. Routinely, we are able to achieve ± 10% reproducibility without difficulty, and with special care, better than ± 5%. The dynamic range of CF-FAB is shown in Figure 3 for the analysis of several solutions of a peptide which contained 0.7 ng (500 fmol), 13.5 ng (10 pmol), and 675 ng (500 pmoles) per 0.5 μL injection. It is seen that even at the high load concentration, tailing of the peak is not excessive and would be acceptable for most applications.

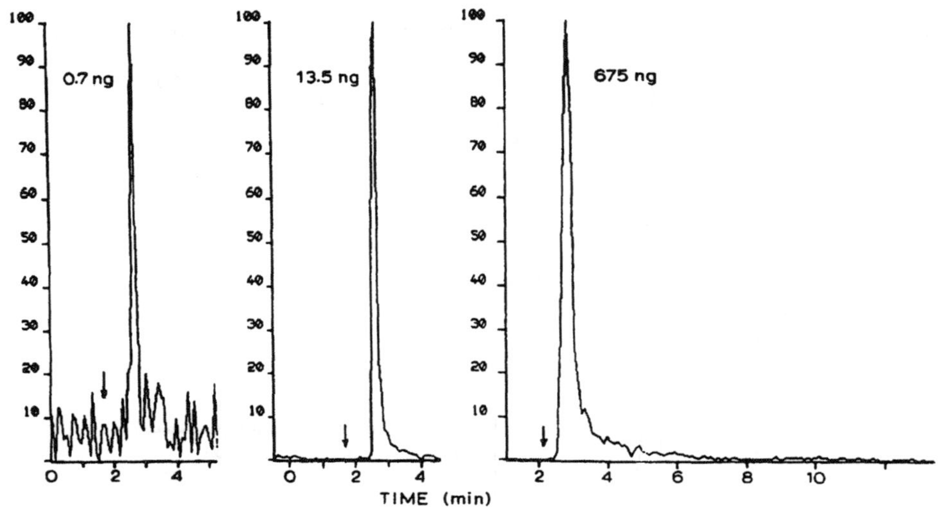

Figure 3. Selected ion chromatograms of the molecular ion region (m/z 1345-1351) for the CF-FAB analysis of various amounts of Substance P. The arrow shows time of injection. (Reproduced with permission from reference 1).

B. STABILITY

Generally, the criteria we have used for 'stable' operation of the CF-FAB probe are; a) obtaining less than ± 10% variation in monitoring of a background ion, for example, m/z 42 for acetonitrile or m/z 185 for glycerol, b) observation of no significant ripple in the peak tops on such ions, c) achieving a reproducibility of peak injection area for a standard compound such as substance P of ± 10% or less, and d) achieving a steady ionization gauge reading of about 2×10^{-4} torr (at flow rates of 5 μL per minute). Under these conditions, the CF-FAB probe is stable and can be used effectively for a variety of different analyses. The CF-FAB probe for the MAT 90 high resolution instrument, shown in Figure 1, has two features which help in achieving stability. First, filter pads are used to remove excess fluid from the target surface, thereby helping to prevent bumping and uneven evaporation which can lead to unstable ion current measurements. Second, the probe is equipped with a mechanical device which moves the capillary in and out of the plane of the target surface, providing a fine position control of the capillary. The ion current stability achieved with this CF-FAB probe is shown in the Figure 4 (3). The top portion of the figure shows the selected ion monitoring of m/z 185 of glycerol, and the bottom of the figure, the total ion current recorded over a period of 70 min. The average fluctuation of the total ion current is less than ± 8%.

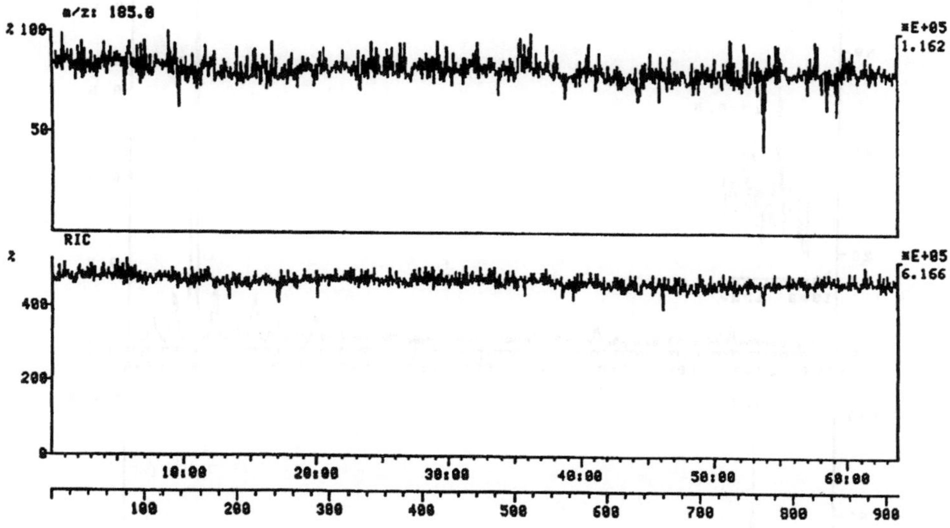

Figure 4. Stability of CF-FAB using a Finnigan MAT 90 high resolution instrument and probe as shown in Figure 1. (Top) Selected ion chromatogram of m/z 185; (Bottom) Total ion chromatogram.

C. MEMORY EFFECT

When the CF-FAB technique is used as an interface for chromatographic methods, it is important to keep memory effects to a minimum. We have evaluated the memory effect by performing a series of analyses with two peptides, substance P ($[M+H]^+$ = 1348) and renin substrate tetradecapeptide ($[M+H]^+$ = 1761), which were alternately injected at a level of 100 pmol each over a one hour period (4). The experiment was performed on a Finnigan MAT 90 which was scanned from m/z 1200-1800. Figure 5 (top) shows the FAB mass spectrum of the last injection of the renin substrate with the $(M+H)^+$ region for substance P expanded in intensity. Only a low ion intensity at m/z 1348 is seen, i.e., less than 0.1% of full scale. Similarly, Figure 5 (bottom) shows the spectrum for the last injection of substance P with the $(M+H)^+$ region for renin substrate peptide expanded in intensity. Again, the memory effect is less than 0.1% of full scale.

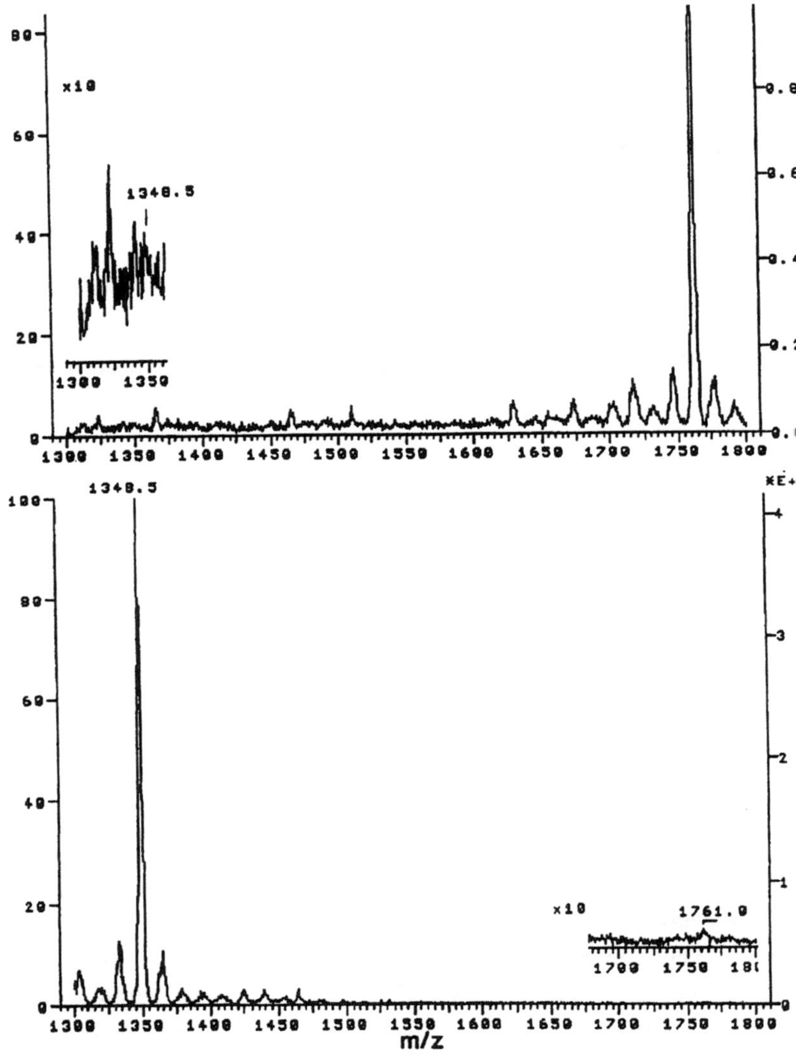

Figure 5. CF-FAB mass spectra of the final injections of 100 pmol each of (top) renin substrate tetradecapeptide and (bottom) Substance P after 20 alternate injections of each peptide. The inset shows a X10 background expansion.

D. PERFORMANCE ADVANTAGES

The reduction of the organic liquid matrix to 5% by volume with the remaining 95% as water leads to some significant performance advantages for CF-FAB relative to standard FAB. First, there is a decrease in the intensity of non-specific background ions. This background or 'peak-at-every-mass' is produced from radiation damage of the glycerol and is significantly reduced when the concentration of organic matrix is reduced, allowing structural features to be seen in the spectrum where they were not readily discernible otherwise.

A second advantage is the increase in sensitivity and limit of detection observed for many compounds analyzed by CF-FAB. Figure 6 shows the analysis of 100 fmol of substance P obtained with standard

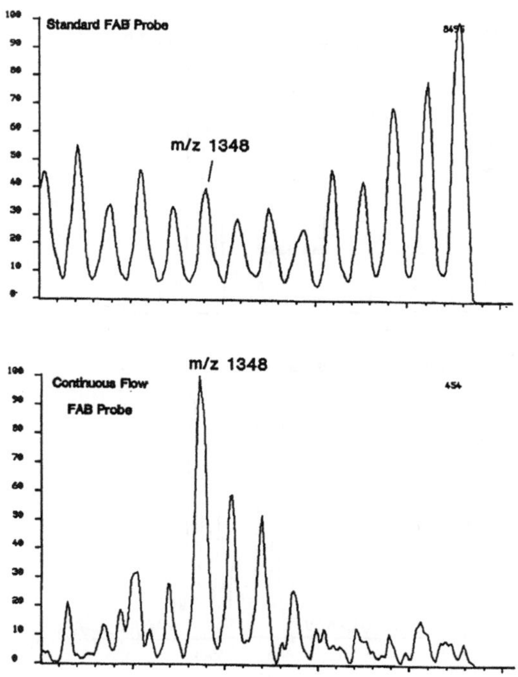

Figure 6. Comparison of FAB mass spectra for the analysis of 100 fmol (136 pg) of substance P taken with (top) standard FAB and (bottom) CF-FAB. (Reproduced with permission from reference 2).

FAB and CF-FAB (2). With standard FAB, the $(M+H)^+$ ion at m/z 1348 is not discernible above background chemical noise as shown in the top portion of the figure, whereas with CF-FAB, the molecular ion is seen at a signal-to-noise level of approximately 5 to 1. It is noted that the number of counts observed in these analyses are remarkably different. The standard FAB spectrum shows a background of 8,495 counts; CF-FAB gives a molecular ion intensity of 454 counts. This example underscores the tremendous decrease in background chemical noise which can be achieved with CF-FAB.

Another performance advantage of CF-FAB involves a decrease in the ion suppression effect which has been described by a number of workers using standard FAB (5, 6). Generally, this effect is the consequence of the inability of certain compounds to occupy the surface layers of the sample droplet and, therefore, little or no ion current from this analyte is produced in the bombardment process. For example, Figure 7 shows the portion of the tryptic digest of

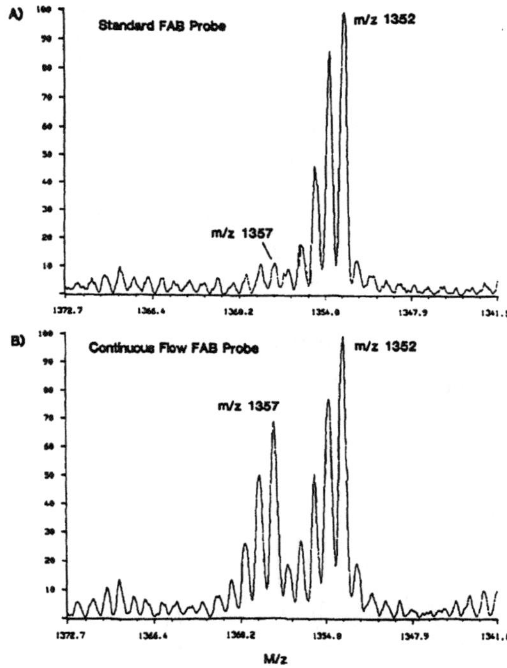

Figure 7. Partial FAB mass spectra of the tryptic digest of glucagon obtained with A) standard FAB, and B) CF-FAB. (Reproduced with permission from reference 6).

glucagon in which there are two peptides which should be present in the bulk sample at equimolar concentrations, one at m/z 1352 and the other at m/z 1357. Using standard FAB, the latter peptide is not readily observed in the spectrum. When the same sample is analyzed by CF-FAB, a significant peak at m/z 1357 is recorded. The diminution or reversal of the ion suppression effect seen here using CF-FAB is most probably the result of thorough and constant mixing of the surface layers of the liquid as the liquid flows over the surface of the target.

III. APPLICATIONS OF CF-FAB

Continuous-flow FAB has been used in a variety of applications involving the analysis of biological compounds in aqueous solutions. These include a) batch sample processing where a large number of analyses are to be performed repetitively for a given compound, b) on-line reaction monitoring where the concentrations of particular compounds may be changing with time, c) in vivo drug pharmacokinetics for the on-line metabolism of specific drugs and drug metabolites, d) direct analysis of biological fluids such as plasma and urine using MS/MS techniques and e) combined liquid chromatography/mass spectrometry in which the CF-FAB probe is used as the interface. Examples will be given to illustrate some of these applications. The use of CF-FAB for LC/MS is presented later in this volume.

On-line reaction monitoring with CF-FAB has been accomplished using two basic procedures; flow-injection and continuous-flow analysis (7, 8). The flow-injection mode basically involves the repetitive injection of small (0.5 μL) aliquots of a reaction solution into a carrier liquid stream flowing at approximately 5 μL per minute into the mass spectrometer. This is shown diagrammatically in the Figure 8. A syringe pump is used to pump the carrier solution through the narrow bore capillary system. Use of a microprocessor controlled syringe pump allows automatic sampling of a reaction reservoir and automatic sample injection for unattended on-line monitoring.

The constant-flow mode, shown diagrammatically in Figure 9, involves continuous pumping of a reaction solution directly into the mass spectrometer. This can be achieved either by using atmospheric pressure in a pressure-regulated system or by use of a syringe pump in a flow regulated system. A small amount (5%) of glycerol is used either in the reaction solution or is added in a make-up tee prior to the sample solution entering the mass spectrometer. Although the atmosphere driven system is mechanically simpler and less expensive to set-up, it is more prone to blockage of the capillary by particulate matter. Applications of these modes of operation will be illustrated with several examples.

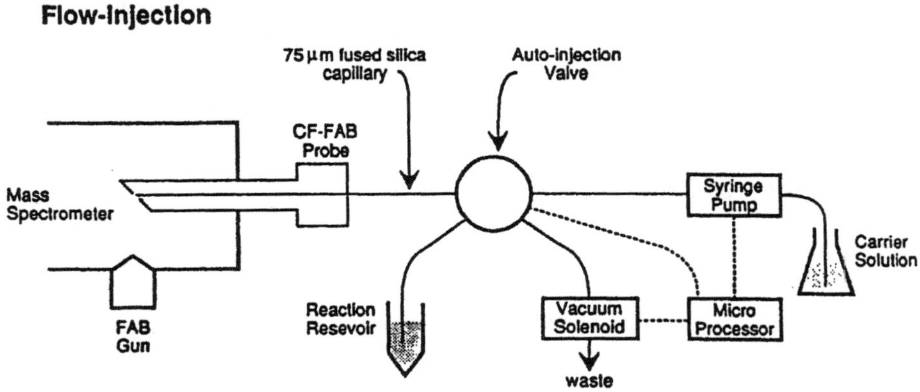

Figure 8. Experimental set-up for the flow-injection analysis of reactions using the CF-FAB interface. (Reproduced with permission from reference 7).

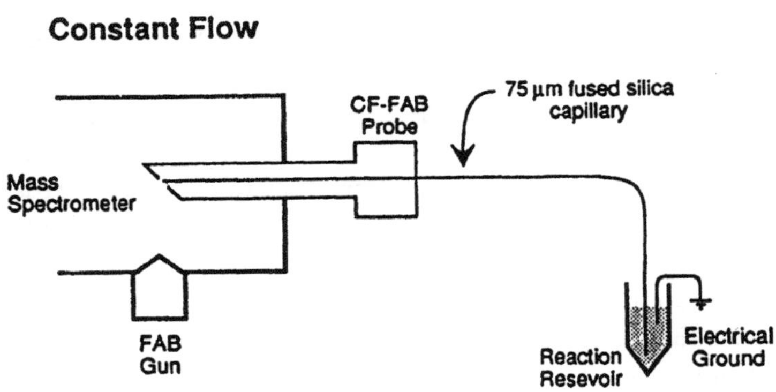

Figure 9. Experimental set-up for the constant-flow monitoring of a chemical reaction using the CF-FAB interface. (Reproduced with permission from reference 7).

A. CONSTANT REACTION MONITORING

The on-line continuous monitoring of enzymic reactions is important for kinetic analysis particularly where multiple products need to be followed. For example, for the hydrolysis of α-MSH by the enzyme pepsin, two fragments are expected, one at m/z 1372 and the other at m/z 1025 (8). Figure 10 shows such an analysis, using selected ion monitoring techniques, for the $(M+H)^+$ of the substrate at m/z 1665, and also those of the two products at m/z 1372 and 1025.

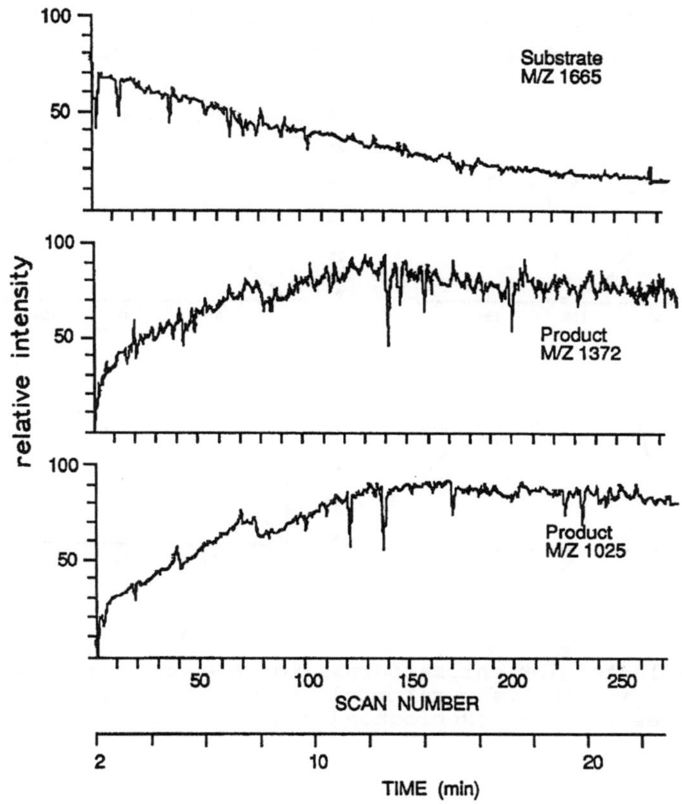

Figure 10. CF-FAB analysis of the hydrolysis of α-MSH (Substrate) by pepsin. The chromatograms obtained for the $M+H^+$ ions of the substrate and two products of the reaction are shown. (Reproduced with permission from reference 8).

From the decrease in the intensity of the molecular ion of the substrate and/or concomitant increases in that of the prodücts, initial rates can be easily calculated for these reaction conditions.

The constant-flow reaction monitoring mode has also been used to follow the action of a mixture of the enzymes carboxypeptidase Y and P in the digestion of ribonuclease S peptide (9). As these enzymes hydrolyze C terminal residues from the peptide, new molecular species are produced from each hydrolysis step. Selected ion chromatograms of these $(M+H)^+$ ions would produce curves which would be expected to rise and fall as these species are produced and they themselves are subsequently hydrolyzed. Several of these ion chromatograms are plotted in Figure 11 for m/z 2095 (M1) representing molecular species of the polypeptide with one C-terminal residue removed, m/z 2025 (M2) for two residues removed, m/z 1937 (M3) for three residues removed, and m/z 1547 (M7) for seven residues removed.

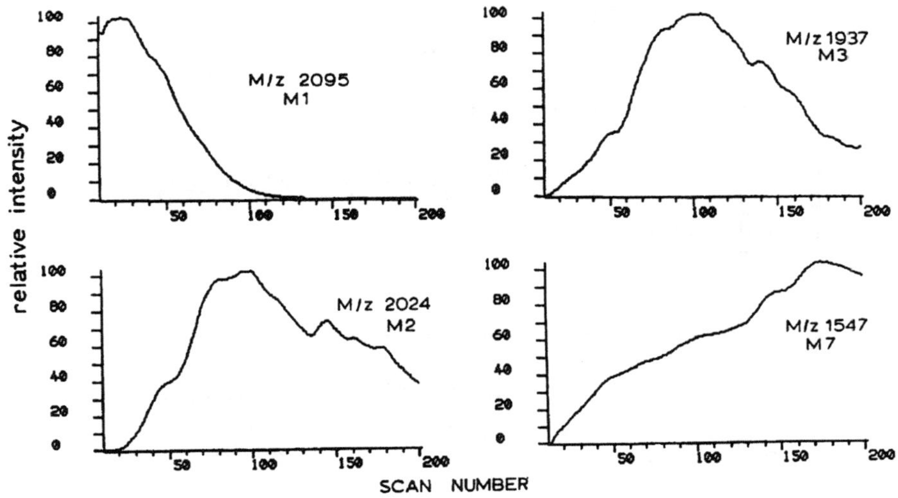

Figure 11. Selected ion chromatograms for the on-line monitoring of the hydrolysis of ribonuclease S peptide by carboxypeptidases Y and P. (Reproduced with permission from reference 9).

The individual FAB mass spectrum at any particular time point can be quite complex, showing a number of molecular species for truncated polypeptides produced in the hydrolysis as well as adduct ions and radiation damage ions for each molecular species.

B. FLOW-INJECTION REACTION ANALYSIS

Operation of the CF-FAB system in the flow-injection mode for on-line analysis of biochemical reactions has also been demonstrated (7, 8). In examples involving the monitoring of multi-reaction processes using a single enzyme, individual compounds in complex mixtures of reactants and products are simultaneously monitored. In one experiment, trypsin was allowed to react with a mixture of 11 peptides, ranging in molecular weights from 1800 to 700 daltons. Each peptide had at least one tryptic hydrolysis site. Figure 12 illustrates the flow-injection analysis for one of these peptides;

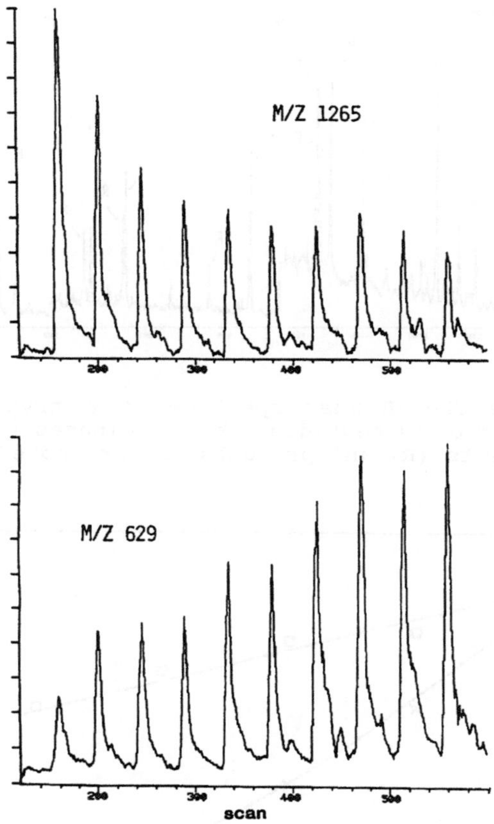

Figure 12. Chromatograms for the $M+H^+$ ions from the time-course tryptic hydrolysis of physalaemin (m/z 1265) and a fragment product (m/z 629) obtained with CF-FAB. (Reproduced with permission from reference 7).

m/z 1265 for the molecular ion of the peptide physalaemin and m/z 629 for one of its tryptic fragments. The decrease in the intensity of the molecular species of this peptide and the increase in that of the product is readily seen. Figure 13 shows a FAB mass spectrum from a single time point in this reaction in which a number of molecular species from the original substrates and also those for peptide products produced in the reaction can be identified. From these time-course data one can calculate the relative rates of specific cleavage reactions in such a mixture. Figure 14 shows reaction rate curves for three peptides hydrolyzed in the reaction in order to show the wide distribution in the rates of hydrolysis which can be monitored in this single experiment.

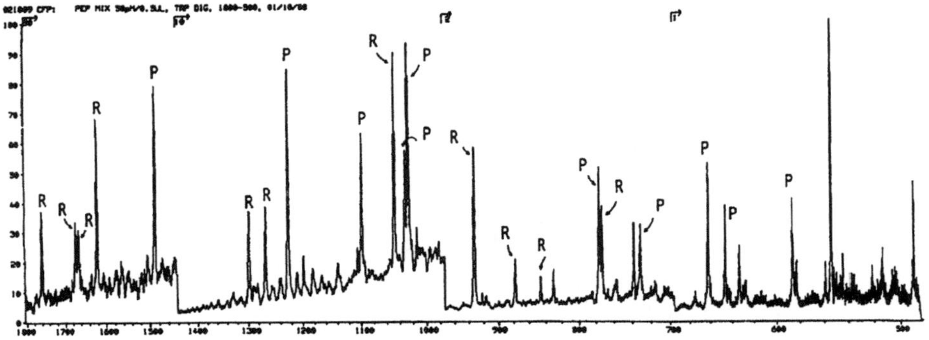

Figure 13. Single CF-FAB mass spectrum of a tryptic digest of a mixture of 11 peptides taken 10 minutes into the reaction. Reactants (R) and products (P) are noted.

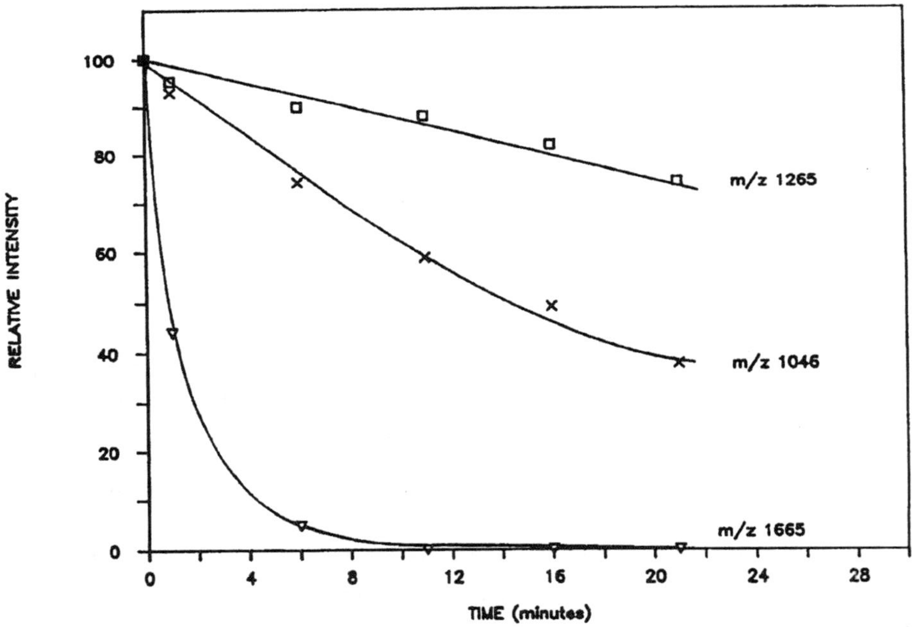

Figure 14. Relative hydrolytic rates obtained from the reaction shown in Figure 13. The curves are for physalaemin (m/z 1265), angiotensin II (m/z 1046), and bombesin (m/z 1665). (Reproduced with permission from reference 7).

The use of multiple enzymes to cleave a single polypeptide substrate to obtain sequence data has also been shown. In this application, both an endopeptidase, subtilisin, and exopeptidases, carboxypeptodiases Y, A and B, were used in a mixture to obtain sequence information from substrate peptides. For example, bombesin was subjected to this enzyme cocktail and the reaction solution analyzed for products (8). Figure 15 diagrams the results which indicates a single internal clip was made in the substrate to give

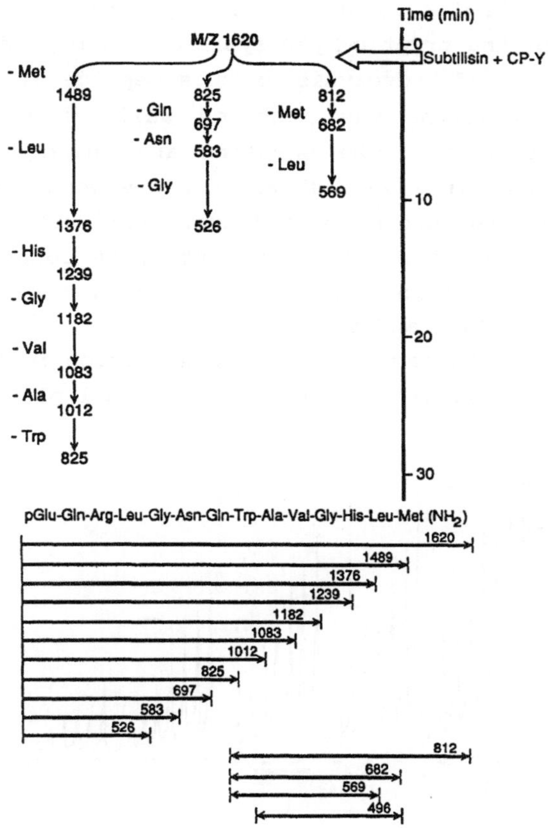

Figure 15. Time-course production of peptides identified by CF-FAB for the on-line monitoring of the hydrolysis of bombesin with a 10:1 mixture of carboxypeptidase Y and subtilisin. (Reproduced with permission from reference 8).

peptide fragments with $(M+H)^+$ ions at m/z 526 and 569. At the same time, C-terminal residues were removed from the intact peptide and the two resulting products. When all the peptides identified were arranged in the order of descending mass, the amino acid sequence of the original sample could be calculated from the difference in masses between successive peptides. A similar analysis of ribonuclease S peptide, which contains 20 amino acid residues, showed that three internal (subtilisin) clips occurred in the original polypeptide and, from the carboxypeptidase data, 15 C-terminal residues were identified in sequence.

Kinetic constants have been obtained for enzymatic hydrolyses using CF-FAB in the flow-injection mode, as shown for the measurement of the K_m (the Michaelis constant) and V_{max} (the maximum velocity of the reaction) for the tryptic hydrolysis of the peptide Met-Arg-Phe-Ala (8). The rate of hydrolysis of this peptide at five different substrate concentrations was measured while keeping the enzyme concentration constant. The sampling and injection process was automated (as shown in Figure 8) and samples were analyzed every 3 minutes. A double-reciprocal plot (Lineweaver-Burke plot) of these rate data where 1/S (where S is the substrate concentration) were plotted against 1/V (where V is the velocity of the reaction) gives an X-intercept of $-1/K_m$ and a Y-intercept $1/V_{max}$. Figure 16A shows the total ion chromatogram for the flow-injection analysis for these hydrolyses, and Figure 16B the double-reciprocal plot of these data.

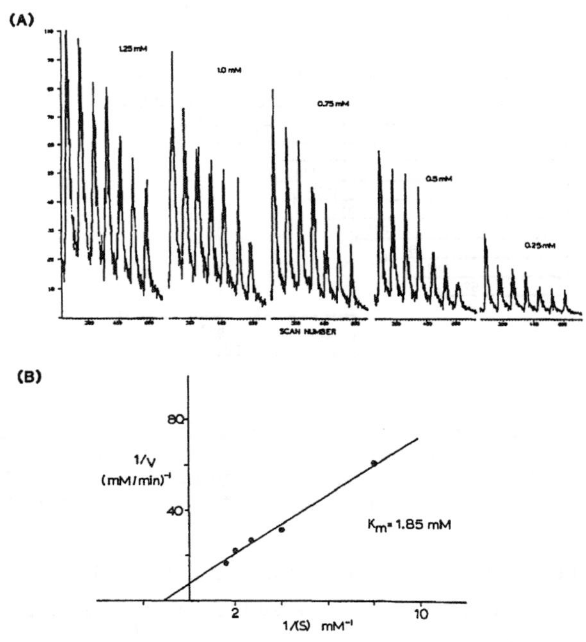

Figure 16. Flow-injection CF-FAB analysis of the hydrolysis of Met-Arg-Phe-Ala by trypsin. See text for description. (Reproduced with permission from reference 8).

The calculated value of K_m for this reaction was 1.85 mM, in close agreement to the value of 1.90 mM obtained previously (10).

A new approach to on-line in vivo drug analyses was recently shown when the CF-FAB interface was used to link a live rat directly to the mass spectrometer for pharmacokinetic studies (11). In this application, a microdialysis probe was inserted into the coronary vein of a rat and this in turn was connected to a reversed phase micro-column for sample concentration. The overall experimental set-up is shown in Figure 17. The microdialysis probe was perfused with

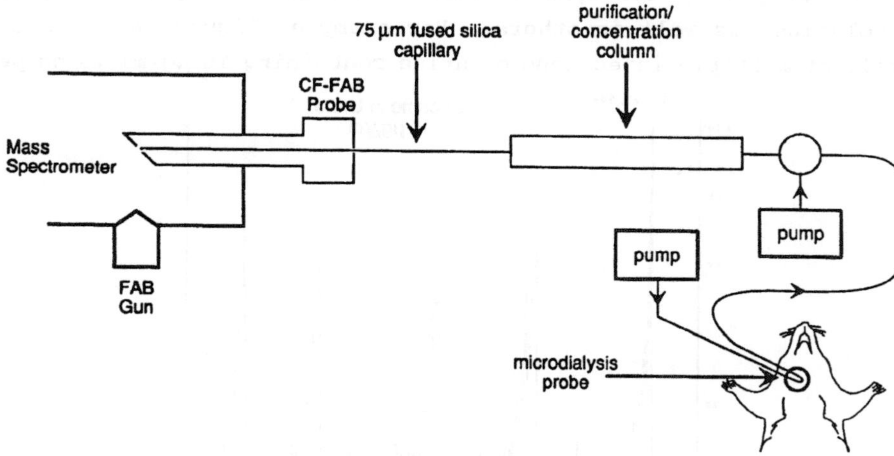

Figure 17. Experimental set-up for the in vivo on-line pharmacokinetic analysis of drugs in the rat. From reference 11.

a carrier solution which deposited the dialysate onto the head of a C-18 reversed phase purification/concentration column. At intervals through the use of a switching valve, the column was eluted with a one-step gradient of acetonitrile and the eluate was allowed to flow directly to the CF-FAB device. A Finnigan MAT TSQ 70 was used in these studies in order to obtain MS/MS spectra because of the high concentration of metabolites in the perfusate. For the example shown here, the rat received an intramuscular injection of penicillin G. The mass spectrometer was operated in the negative ion mode with the first stage set to pass m/z 333, the $(M-H)^-$ molecular species of this drug. The MS/MS capability was essential in monitoring the drug in the complex dialysate because the bombardment process produced virtually a peak at every mass in the spectrum. The collision activated decomposition product at m/z 192, however, gave an intense signal which was derived specifically from the drug. The pharmacokinetic curve generated in this manner from the injection of penicillin G into the test animal showed a maximum concentration at approximately 1.5 hours, a value in agreement with that previously obtained under similar conditions. This application represents perhaps one of the most demanding tests of the CF-FAB device for the detection of drugs in blood dialysate and demonstrates the value of MS/MS to this type of application.

We have also used CF-FAB with MS/MS to directly examine urine for drugs and drug metabolites (12). These drugs included cocaine, codeine, morphine, benzoylecognine, pentobarbital, phenobarbital, phentolamine, as well as others. For example, Figure 18 shows the profile of multiple injections of urine containing 10 μg/ml (5 ng per

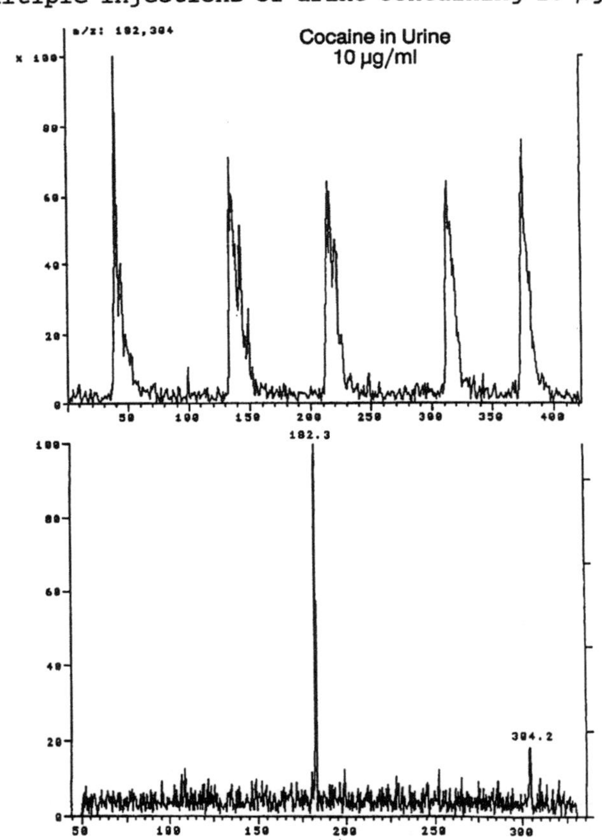

Figure 18. Direct analysis of cocaine in urine with CF-FAB and MS/MS. (Top) Replicate injections of urine spiked with cocaine at 10 μg/mL, (bottom) CAD FAB spectrum from a direct urine injection. From reference 12.

injection) of cocaine. The bottom panel of the figure shows the daughter ion spectrum of cocaine in urine consisting of a surviving parent at m/z 304 and a relatively intense daughter ion at m/z 182. The standard curve was found to be linear from 25 pg. up to approximately 15 ng. Similar results were obtained for the other drugs tested and showed that these techniques could be effectively used on crude and complex biological fluids.

IV. CONCLUSIONS

On-line CF-FAB represents a powerful technique to the biochemist and molecular biologist for the analysis of polar compounds contained in aqueous solutions. These solutions need not be pure and, in fact, may contain rather complex mixtures. The advantages of the on-line CF-FAB experiment are several in that it; a) effectively adds the time dimension to reaction analysis by FAB mass spectrometry, b) provides real-time data and the flexibility to change reaction conditions as the reaction proceeds, c) gives good reproducibility without the need for an internal standard, and d) allows direct and automated sampling of aqueous solution systems. This technique also has some disadvantages; a) the set-up and stability time required generally is on the order of 10 to 15 minutes, b) capillary plumbing connections are critical and low dead volumes need to be scrupulously avoided, and c) source contamination can occur with complex biological fluids and may be troublesome for instruments which do not have removable ion volumes.

In providing a means for the direct analysis of aqueous samples, mass spectrometry more completely enters the domain of the biochemist. There is little doubt that it will help uncover new and exciting aspects of the structure and composition of compounds which are part of the complexity of living cells.

Acknowledgements

The author thanks Millie Martin in obtaining some of the CF-FAB data for reaction monitoring, Shen-Nan Lin for the in vivo pharmacokinetic data, William Seifert for the urinary drug analysis data, and Reinhold Pesch for some of the CF-FAB probe data from the MAT 90.

References

1 Caprioli RM, Fan T, Cottrell JS (1986) Anal Chem 58:2949
2 Caprioli RM, Fan T (1986) BBRC 141:1058
3 Pesch R, Caprioli RM, unpublished results
4 Caprioli RM (1988) Proceedings 36th ASMS Conf on Mass Spectrom, June 5-10, San Francisco 529
5 Naylor S, Findeis F, Gibson BW, Williams DH (1986) J Amer Chem Soc 108:6359
6 Caprioli RM, Moore WT, Fan T (1987) Rapid Commun Mass Spectrom 1:15
7 Caprioli RM (1988) Trends Anal Chem (in press)
8 Caprioli RM (1988) Biochem 27:513
9 Caprioli RM (1987) Mass Spectrom Rev 6:237
10 Caprioli RM, Smith LA (1986) Anal Chem 58:1080
11 Lin SN, Caprioli RM (1988) Proceedings 36th ASMS Conf on Mass Spectrom, June 5-10, San Francisco 1000
12 Seifert WE, Lin SN, Moore, WT, Ballatore A, Caprioli RM (1988) Proceedings 36th ASMS Conf on Mass Spectrom, June 5-10, San Francisco 1321

COUPLING CHROMATOGRAPHIC TECHNIQUES WITH FABMS FOR THE STRUCTURAL ANALYSIS OF BIOLOGICAL COMPOUNDS

Richard M. Caprioli

The Analytical Chemistry Center
and Department of Biochemistry and Molecular Biology
The University of Texas Medical School at Houston
Houston, Texas 77030

I. INTRODUCTION

High performance liquid chromatography (HPLC) has become an indispensable tool to the biochemist and molecular biologist for the separation of polar and charged compounds derived from biological systems. It is obvious that coupling this technique to mass spectrometry provides a powerful mass specific detection and analysis system which is unmatched in terms of its combined specificity and sensitivity. The success of FAB for the analysis of polar compounds has led to a number of approaches for coupling HPLC and FABMS. This article will discuss interfacing the continuous-flow fast atom bombardment (CF-FAB) technique with HPLC (1, 2) and applications of this technique. It is noted that other methods of coupling these have been described, namely, the moving belt (3) and frit-FAB techniques (4). Other types of separation processes have been coupled with FAB as well, including thin layer chromatography (TLC) (5) and capillary zone electrophoresis (CZE) (6). All three separation techniques play significant roles in the biochemistry laboratory, and are significantly enhanced when coupled with mass spectrometry.

II. CF-FAB INTERFACE

The use of the CF-FAB technique as an interface for microbore HPLC is shown diagrammatically in Figure 1. In this set-up, the microbore HPLC used was an Applied Biosystems Model 130A, with some modification, and the mass spectrometer was a Finnigan Mat 90 high resolution mass spectrometer. We have found that a flow rate of about 25 μL/min through the microbore column in such an instrumental arrangement provides good chromatographic performance. For gradient

U. P. Schlunegger (Ed.)
Biologically Active Molecules

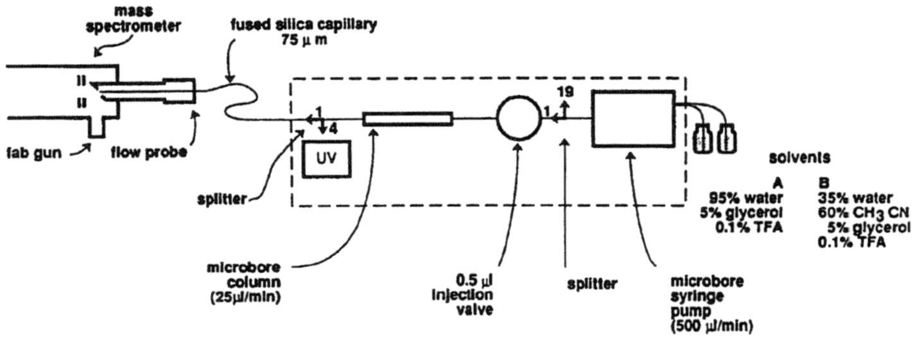

Figure 1. Instrumental set-up for microbore LC/CF-FAB. (Reproduced with permission from reference 2).

elution chromatography, we have installed a 9:1 splitter in front of the injection valve so that gradients can be formed at relatively high flow rates while still allowing approximately 25 µL/min to flow through the column. For peptide analyses, for example, gradients are formed at approximately 250 µL/min with buffer A consisting of 95% water, 5% acetonitrile and 0.1% TFA, and buffer B, 35% water, 60% acetonitrile, 5% glycerol, and 0.1% TFA. We have also used a 4:1 splitter after the microbore column, allowing 5 µL/min to enter the mass spectrometer and 20 µL/min to flow to a UV detector. The split to the UV detector serves two purposes; it provides a UV absorption tracing of the compounds eluting from the column and also allows for the collection of fractions for other analyses and reaction assays. The effect of chromatographic operating parameters such as flow rates, times of gradient formation, column sizes, etc., will be discussed in some detail later.

III. ANALYSIS OF PEPTIDES

One of the major applications of microbore LC/FABMS is the structural analysis of polypeptides and proteins. The general approach first requires proteolytic digestion of the polypeptide with one or more proteases such as trypsin, chymotrypsin, subtilisin, V8 protease or other enzyme. Direct injection of the digest is then made onto the LC column. Gradient elution techniques are used to effect rapid separation of the peptide mixture, usually less than 30 min duration, followed by the mass specific analysis of the eluting compounds. This procedure is useful for many aspects of protein analysis, including (1) verification of the sequence of an active polypeptide where this sequence has been predicted from the DNA analysis, (2) identification of the nature and position of insertions or deletions in the sequence, (3) localization and identification of post-translational modifications such as the addition of oligosaccharides, phosphate, sulfate and acyl groups, and (4) assessment of the purity of protein and polypeptide preparations.

A. SENSITIVITY

The overall sensitivity of LC/FABMS will vary greatly, depending on the type of compounds analyzed and the complexity of the sample. However, several general comments can be made. As a rough estimate at present, in the mass range between 500 and 3000 amu, a sample of 20 pmols of peptide gives approximately 50,000 counts with a background of several thousand counts. The practical lower limit of detection of peptides in enzyme digests is approximately 1-3 pmols giving a signal-to-background chemical noise ratio of about 5:1. In terms of the lower limit of the mass range, although the mass spectrometer is capable of scanning to very low m/z values, practically, the chemical noise below m/z 300 or so generated by FAB tends to limit its use to samples which contain relatively large (nmol) amounts of peptide or other compounds. At the high mass range, small proteins of m/z 6000 can be effectively identified at amounts of 200-300 pmol. This high mass sensitivity does not represent an instrumental limitation other than that imposed by current FAB gun and detector technology. For example, we have successfully separated bovine, ovine, equine, and porcine insulins from a mixture containing 400 pmol of each protein. Using low resolution techniques and peak centroid calculations, the molecular species of each protein could be identified to approximately ± 0.5 daltons. As new ionization techniques and improvements occur, it is expected that the mass range and sensitivity of the technique will similarly increase.

B. CHROMATOGRAPHIC PARAMETERS

The optimization of chromatographic parameters is very important in the analysis of mixtures. Figure 2 shows the effect of three different solvent gradient programs on the separation of peptides in the tryptic digest of horse heart cytochrome C with three microbore columns of different dimensions. The gradient programs used for these studies are as follows; with solvent A = 0/5/95/0.1, and solvent B = 60/5/35/0.1 of acetonitrile/glycerol/water/TFA (V/V).

Program I: 5% B, hold 0.1 min; 5-40% B in 2.4 min; 40-60% B in 1.0 min; 60% B thereafter.

Program II: 5% B, hold 0.1 min; 5-40% B in 4.9 min; 40-60% B in 1.0 min; 60% B thereafter.

Program III: 5% B, hold 0.1 min; 5-40% B in 9.9 min; 40-60% B in 1.0 min; 60% B thereafter.

Examination of the chromatographic performances shows that the chromatogram shown in panel A of Figure 2 represents a rapid analysis (approximately 5 min) but one which shows poor resolution. The chromatogram in panel I represents opposite characteristics, i.e., a longer elution time (about 15 min) and much improved resolution. One can look at the variation in these chromatograms not as

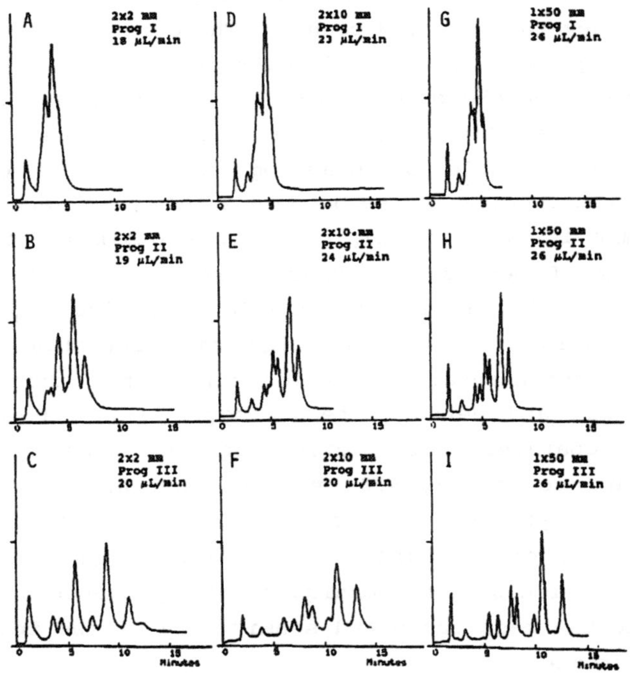

Figure 2. Effect of solvent gradient conditions and column dimensions on resolution of peptides in the tryptic digest of cytochrome c. (Reproduced with permission from reference 2).

limitations but the result of an extremely flexible system that can be tailored to achieve the ends of a particular analysis. If, for example, the mixture is relatively simple and the analytical requirement is to separate non-peptide components such as salts, then a fast analysis may be preferred. This may also be the case when on-line monitoring of a system or process is being accomplished where the time between repetitive analyses is quite short. On the other hand, for complex mixtures, resolution of the peaks in the chromatogram and the limitation of the number of peptides entering the mass spectrometer at any particular time point may be the primary chromatographic goal. For still further improved resolution, longer columns such as a 250 mm column may be used, although this will increase the elution time to about 30-40 min or more under the conditions outlined above. Thus, there is a basic trade-off between overall analysis time and resolution in such a system.

The effect of the flow rate of the eluting buffer passing through the microbore columns is important in this instrumental arrangement. Using three different chromatographic columns (2x2 mm, 2x10 mm, and 1x50 mm), it was found that the optimal flow rate was approximately 25 μL/min. Lower flow rates significantly compromised resolution and gave long analysis times. Although higher flow rates increased the resolution of the column somewhat, the effect was not significant in comparison to the drawback of the need for greater sample splitting after the column to give the 5-10 μL/min required by the mass spectrometer. This 10 μL/min flow rate is a consequence of the limitation of the pumping capacity of commercial mass spectrometers. Although larger pumps may be installed on these instruments, we have not found this to be necessary to obtain good LC/MS data.

Perhaps one of the most important operating parameters is the time required for the production of the gradient. This is clearly seen in panels G, H, and I in Figure 2 for the three different gradient programs. The 1x50 mm microbore column was used in this study with a flow rate of 26 μL/min. The steepness of the gradient varied from 5-40% B over 2.5 min in panel G, over 5 min in panel H, and over 10 min in Panel I. Again the time/resolution trade-off is clearly seen and may be used to advantage by the investigator to fit his particular analytical need.

C. APPLICATIONS

The LC/CF-FAB analysis of the tryptic digest of β-lactoglobulin A is shown in Figure 3. The mass spectrometer was scanned between m/z 400 to 2700 in 30 sec. Approximately 200 pmols of the digest were loaded onto the column with 40 pmols entering the mass spectrometer after separation and a post-column split. Figure 4

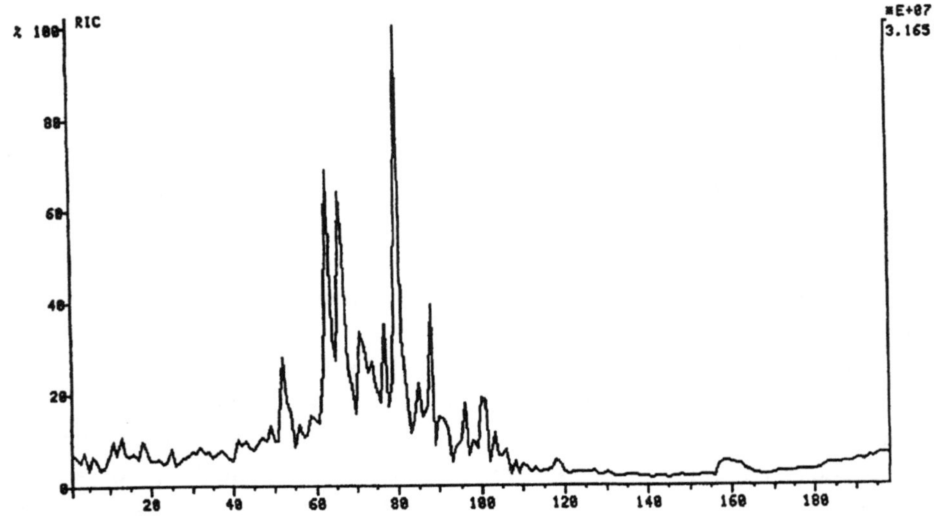

Figure 3. Total ion chromatogram from the mass spectrometric analysis of 50 pmol of the tryptic digest of β-lactoglobulin A.

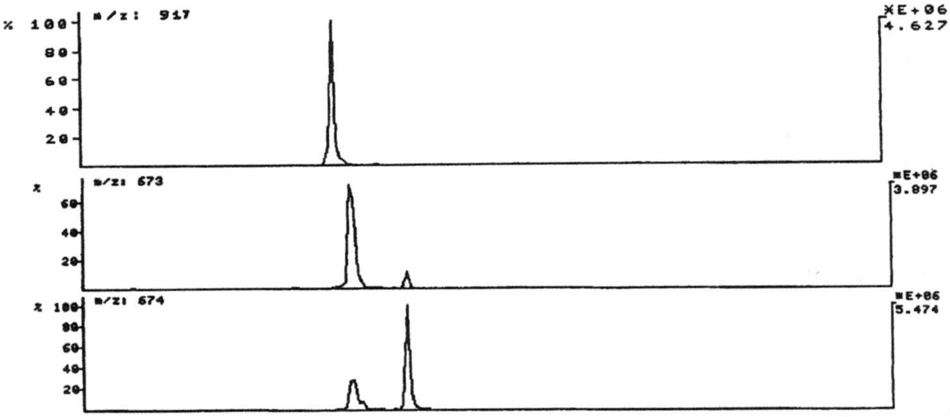

Figure 4. Selected ion chromatograms of several peptides produced in the tryptic digest of β-lactoglobulin A.

demonstrates the use of selected ion chromatograms for identifying specific peptides in the mixture. The top chromatogram shows the SIC for m/z 917 corresponding to $(M+H)^+$ for the peptide Leu-Asp-Ala-Ile-Asn-Glu-Asn-Lys. The value of chromatographic resolution is illustrated in this example by the detection of two peptides containing mass 674. For one peptide, this represents the $(M+H)^+$ of the monoisotopic species and for the other, the $(M+H)^+$ for the first C^{13} containing species. The identification of the two peptides is clear since they are separated by approximately 6 min in the chromatogram.

In the case of the analysis of proteins of unknown sequence or where a post-translation modification will produce a tryptic peptide of unknown mass, it is extremely useful to produce a 'mass summary' of the LC/MS run in order to identify peptides present. This data presentation is merely a summation of all of the mass spectra in the chromatographic profile, as shown for the β-lactoglobulin A digest sample in Figure 5. Although signal intensity is somewhat decreased

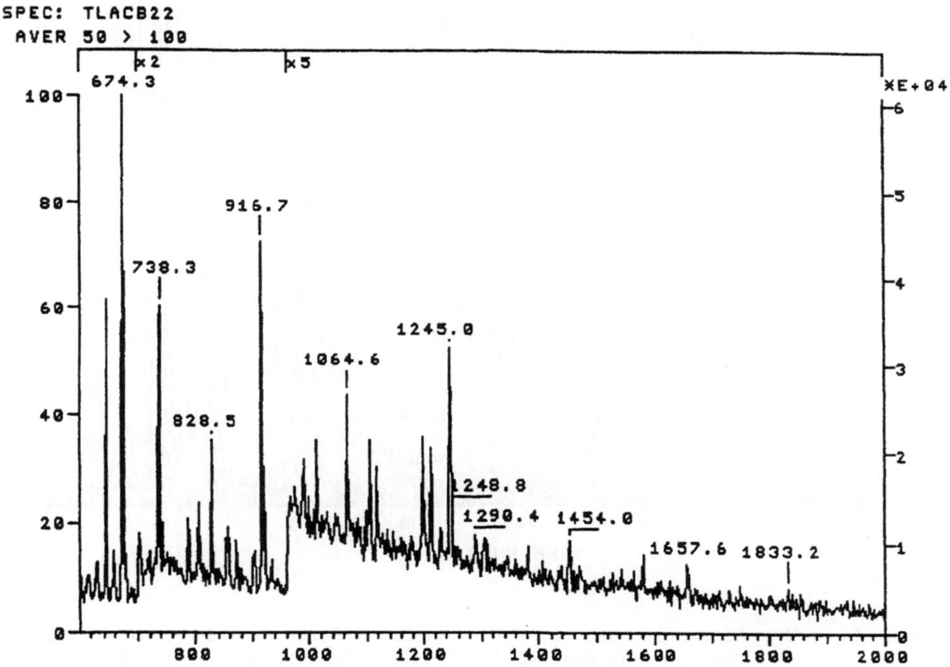

Figure 5. "Mass Summary" of molecular species recorded in the LC/MS analysis of the tryptic digest of β-lactoglobulin A.

in such a summation profile, it allows one to easily identify the peptides present in the analysis. Each peptide can then be searched for individually to determine its actual mass and precise location in the chromatogram.

Another useful presentation of the data is shown in Figure 6 and is essentially a 'peptide chromatogram'. This is accomplished by overlaying independently normalized selected ion chromatograms for each peptide found in the mixture and represents a clear way of recording the positions of the peptides in the total ion chromatogram. In all, 19 peptides which fell within the mass range scanned were identified in the analysis of the tryptic digest of β-lactoglobulin A.

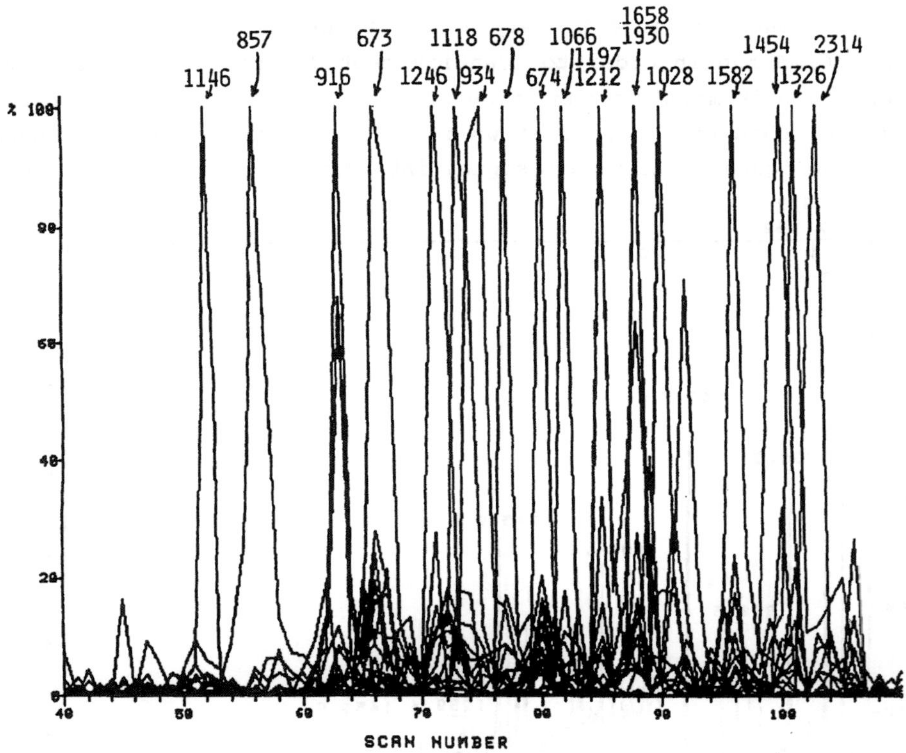

Figure 6. "Peptide Chromatogram" obtained from overlaying selected ion chromatograms of individual peptides identified from the analysis of the tryptic digest of β-lactoglobulin A.

The LC/CF-FAB analysis of 240 pmol of the tryptic digest of apolipoprotein A-1 (apo A1, approximate mw 23,000 daltons) led to the identification of over 85 individual peptides, although only about 30 were expected (7). Since the lower limit of detection of this system is approximately 1-2 pmols for such complex digests, it was clear that protein contaminants in this preparation at the 1-2% level were being detected. To expand further on the issue of sensitivity,

Figure 7A shows the comparison of the total ion chromatograms generated for the analyses of 240 pmols, 24 pmols, and 2.4 pmols of this digest, and Figure 7B, the mass spectrum of one of the fragment peptides at m/z 1013. It is seen that even at a level of several

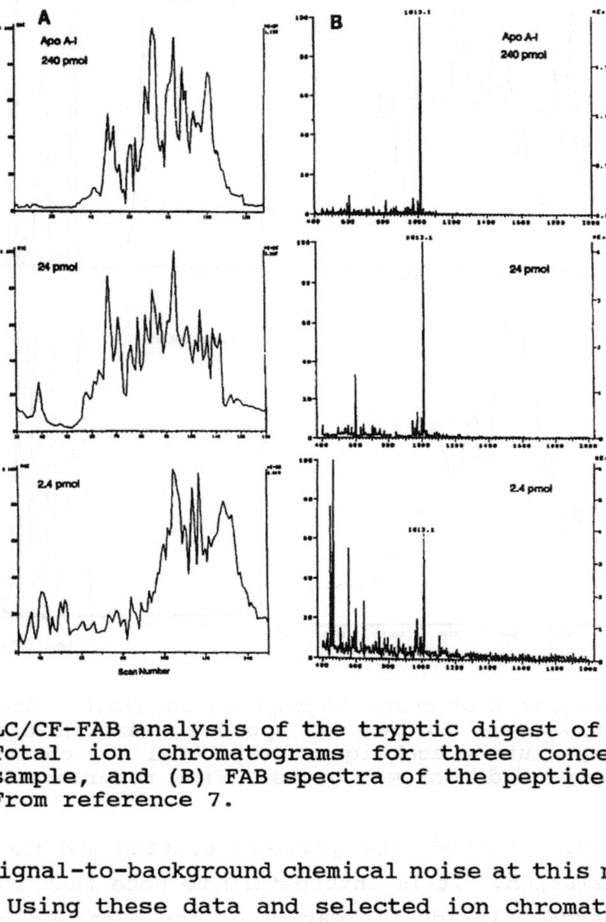

Figure 7. LC/CF-FAB analysis of the tryptic digest of apo A-1. (A) Total ion chromatograms for three concentrations of sample, and (B) FAB spectra of the peptide at m/z 1013. From reference 7.

pmols, the signal-to-background chemical noise at this mass is better than 10:1. Using these data and selected ion chromatograms, 23 of the tryptic peptides from apo A1 expected in this mass range were identified.

Often it is desirable to achieve very rapid analysis and sacrifice resolution for a specific analytical purpose. Figure 8A shows the peptide chromatograms for the apo A-1 digest separated under two different conditions. The top chromatogram was obtained from the gradient elution and the bottom chromatogram from a one-step elution of 24 pmols of the digest. The one-step elution involves changing from 0% B to 100% B within 1 min from the time of sample injection. For simplicity, Figure 8B shows some selected peptides chosen throughout the chromatographic range. Again, the top

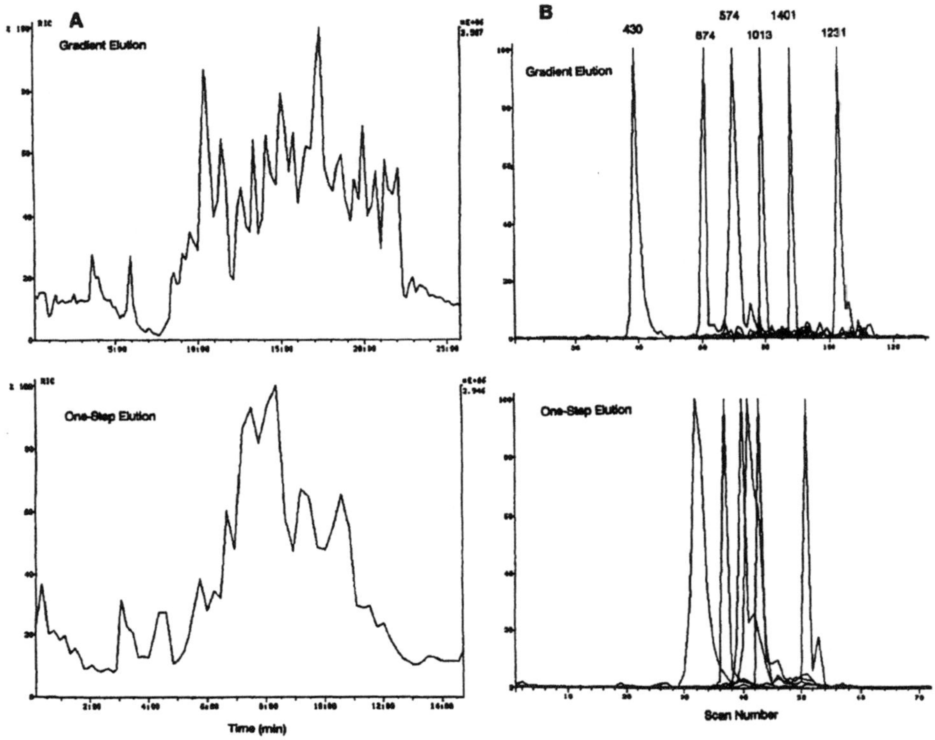

Figure 8. Comparison of chromatographic resolution obtained for the analysis of 24 pmol of apo A-1 using gradient and one-step elution techniques. (A) Total ion chromatograms and (B) peptide chromatograms. From reference 7.

chromatogram was obtained from gradient elution and the bottom from the one-step elution. It is interesting to note that in both cases, the mass spectrometric identification of the peptides was made with equal facility even though the chromatographic resolution was markedly different. However for more complex mixtures where greater numbers of compounds could elute in close proximity and where mass redundancies may occur, gradient elution would be the preferred method.

The identification of modifications and chemical changes in the side chains of polypeptides are of considerable interest particularly in the area of recombinant DNA protein technology. A number of investigators have used LC and MS techniques separately for the analysis of proteins that have undergone post-ribosomal modification reactions. These include truncation of the N- and C- termini, disulfide bridging, addition of side chain sugar, phosphate, or sulfate groups, N-acylation reactions, γ-carboxylation and oxidation of side chains, etc.

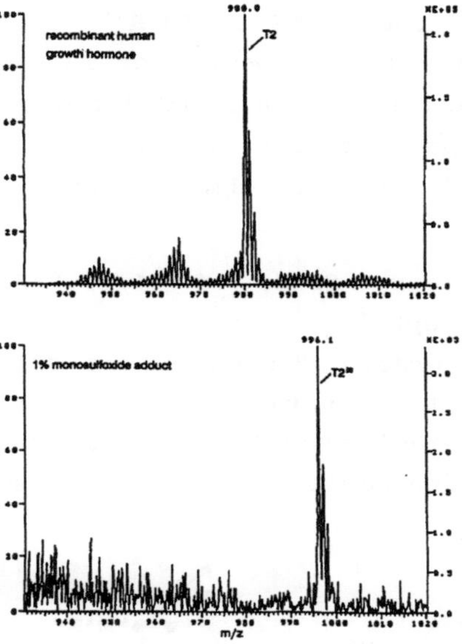

Figure 9. FAB mass spectra of the T2 (tryptic) peptide from recombinant growth hormone. (Top) Molecular region of the expected T2 peptide; (Bottom) Molecular region of the oxidized T2 peptide. From reference 7.

The use of LC/CF-FAB in such studies can be illustrated by the analysis of recombinant human growth hormone in which the identification of the location of the oxidation of a side chain was of interest. To demonstrate the capabilities of LC/MS, a test sample was prepared in which a purified preparation of hormone was deliberately doped with a 1% monosulfoxide adduct of the hormone. (The sample was supplied courtesy of Jerry Becker, Eli Lilly Co., Indianapolis, USA). The LC/CF-FAB analysis of the tryptic digest of this hormone showed that in addition to the tryptic fragments expected, an additional fragment (termed T2*) was also identified. This fragment had an m/z value for the $(M+H)^+$ ion which was 16 mass units higher in mass than that of T2. This mass difference indicates the addition of an oxygen atom to this peptide, presumably on the sulphur atom of either a methionyl or cysteinyl peptide. This is readily seen from the mass spectra shown in Figure 9 for T2 and T2*. The T2 tryptic fragment is unique in the mixture and is derived from positions 9 to 16 of the original hormone polypeptide chain. The sequence of T2 is Leu-Phe-Asp-Asn-Ala-Met-Leu-Arg. Amino Acid analysis of this fragment conclusively proved the presence of methionine sulfoxide. It should also be noted from Figure 9 that the number of ion counts for m/z 996 ($[M+H]^+$ for T2*) appears to be approximately 1.4% of m/z 980 ($[M+H]^+$ for T2). This example quite nicely shows the capability of this LC/MS technique to identify and give a good quantitative estimate of the level of impurities at the 1% contamination level and at an absolute amount of impurity of 1-2 pmol.

Although the primary focus of LC/CF-FAB has been on the identification of peptides, work has also been done with oligosaccharides mixtures. Boulenguer et al. (8) have recently published results on the separation and analysis of permethylated oligosaccharides obtained from hen oviduct ovomucoid. A slurry packed C-18 fused silica capillary column was used. Figure 10 shows

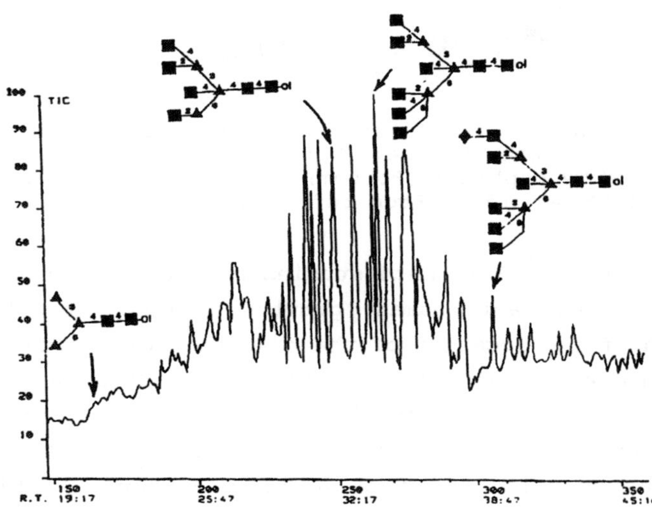

Figure 10. Total ion chromatogram from the LC/CF-FAB analysis of permethylated oligosaccharides obtained using a packed fused-silica microcolumn. (Reproduced with permission from reference 8).

the total ion chromatogram from this analyses with several of the oligosaccharides identified on the chromatogram using a Kratos MS50 high resolution mass spectrometer with CF-FAB. Their results from the analysis of permethylated oligosaccharides showed that excellent chromatographic separation and mass spectrometric identification can be done on about 100 ng (about 100 pmol of a derivatized tetramer or pentamer), with a lower limit of detection of about 10 pmol.

D. IMMUNOAFFINITY CHROMATOGRAPHY

Immunoaffinity chromatographic techniques have been used successfully with the CF-FAB interface to analyses endogenous compounds. In one such application by Lisek et al. (9), both immunoaffinity chromatography and HPLC have been used for the quantitative analysis of the endogenous neuropeptide substance P. In their procedure, the biological sample was obtained in aqueous solution and a deuterated internal standard of substance P was added. After extraction, the sample was passed through an immunoaffinity chromatography column to purify the neuropeptides and the sample was then split. One portion was used for a radioimmunoassay for substance P and the other portion subjected to on-line HPLC CF-FAB. Figure 11 shows the micro column LC/CF-FAB selected ion monitoring

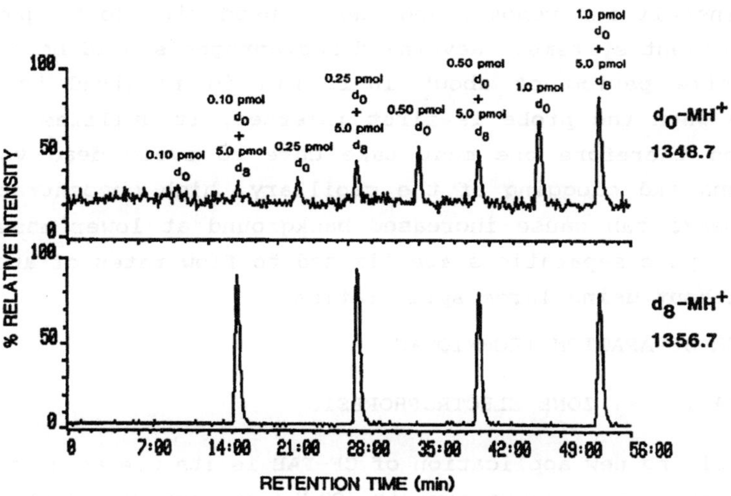

Figure 11. Quantitation of Substance P with microcolumn LC/CF-FAB showing sample injections containing increasing amounts of Substance P-d_0 alternating with the same sample spiked with the d_8 internal standard. From reference 9.

for substance P (d_0) and for internal standard (d_8) obtained from the injection of various amounts of the d_0 species (0.1-1.0 pmol) with alternate injections containing additions of 5 pmol of the d_8 species. It is seen from the Figure that the level of detection of the d_0 species is approximately 0.1 - 0.2 pmols. The lower channel shows that for injections containing only the d_0 species, no signal is recorded on the d_8 species channel. These data were then used to establish a quantitative response curve. Determination of endogenous substance P in biological samples were made by measuring the d_0/d_8 ratio for this peptide and calculation of the amount present from the standard curve. For example, an 8 mg. sample of cat spinal cord was found to contain 39.0 pmol of substance P by this method, in good agreement with a radioimmune assay value of 40.7 pmol.

E. SUMMARY

The CF-FAB interface can be used effectively for the analyses of complex mixtures separated by LC. It is compatible with high efficiency microbore techniques and has shown high sensitivity and low background, making full use of the relatively high concentration of eluting compounds obtained with this LC method. It is compatible with techniques used for high mass analysis and can be used in magnetic mass spectrometers having high voltage ion sources. It is easy to install and remove, and can be used with both aqueous and organic solvent systems. Several disadvantages should be noted; an initial time period of about 10-15 min is required to achieve stability when the probe is first inserted, it utilizes capillary tubing and therefore one must take care to avoid dead volume in connections and clogging of the capillary, high concentrations of eluting peaks can cause increased background at lower masses, and chromatographic separations are limited to flow rates of about 5-25 μL/min without using large split ratios.

IV. OTHER SEPARATION TECHNIQUES

A. CAPILLARY ZONE ELECTROPHORESIS

An exciting new application of CF-FAB is its use as an interface for capillary zone electrophoresis (CZE)/mass spectrometry)10, 11). Minard et al. (10) have recently demonstrated the feasibility of such an instrument arrangement as shown in Figure 12. One of the major problems encountered in coupling CZE and mass spectrometry is the fact that the terminus of the capillary inside the mass spectrometer is at high vacuum and therefore draws liquid into the

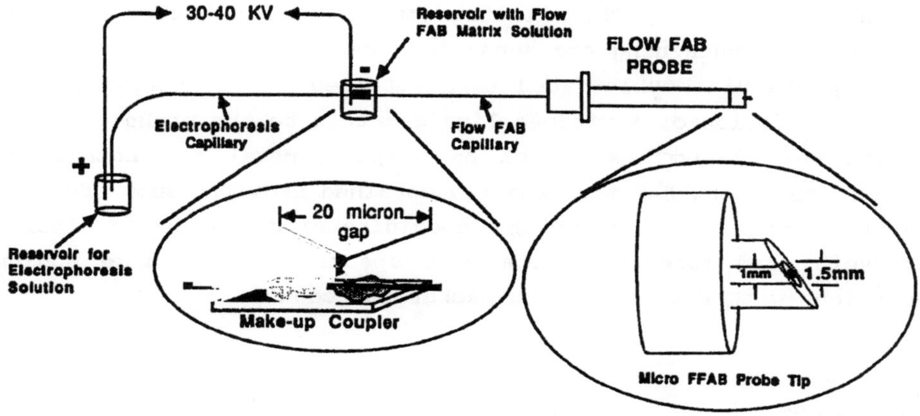

Figure 12. CZE/MS set-up using the CF-FAB interface. From reference 10.

spectrometer. However, optimal separations by CZE occur when there is no mechanical force causing net flow of liquid, only electroosmotic flow which gives flow rates of a hundred nanoliters or so per minute from the anode to the cathode. These investigators have minimized this difficulty by allowing a 20 micron gap in the capillary at the site of the cathode; excess liquid is pulled in from the cathode solution to the mass spectrometer as well as the separated compounds emerging from the end of the CZE capillary (see Figure 12). The authors illustrated this technique with the analysis of leucine enkephalin. An intense $(M+H)^+$ at m/z 556 was observed, with low background.

Other investigators have reported the interfacing of CZE and mass spectrometry, although not with FAB. Smith et al. (6) have described an interface which consists of an electrospray ion source and a quadrupole mass filter and has shown excellent separations and mass detection. Although other methods of interfacing are being investigated, it is clear that CZE/MS will provide a powerful tool for the biochemist and molecular biologist.

B. THIN LAYER CHROMATOGRAPHY

Thin layer chromatography (TLC) has long been an extremely effective method for the separation of compounds of biological interest. Several investigators have recently reported results in the use of FAB for the mass specific detection of compounds which have been separated by TLC. In the work of Chang et al. (12), the solid support containing the separated compound was transferred to the FAB probe via double-faced tape attached to the sample probe. After the addition of a viscous liquid matrix to the probe tip, the material was placed inside the mass spectrometer and bombarded. Figure 13 shows the FAB mass spectra obtained from TLC spots for the separation of a mixture of the coccidiostatic agents lasalocid, septamycin, and monensin. The mass spectra show intense sodium adduct ions with relatively low background counts.

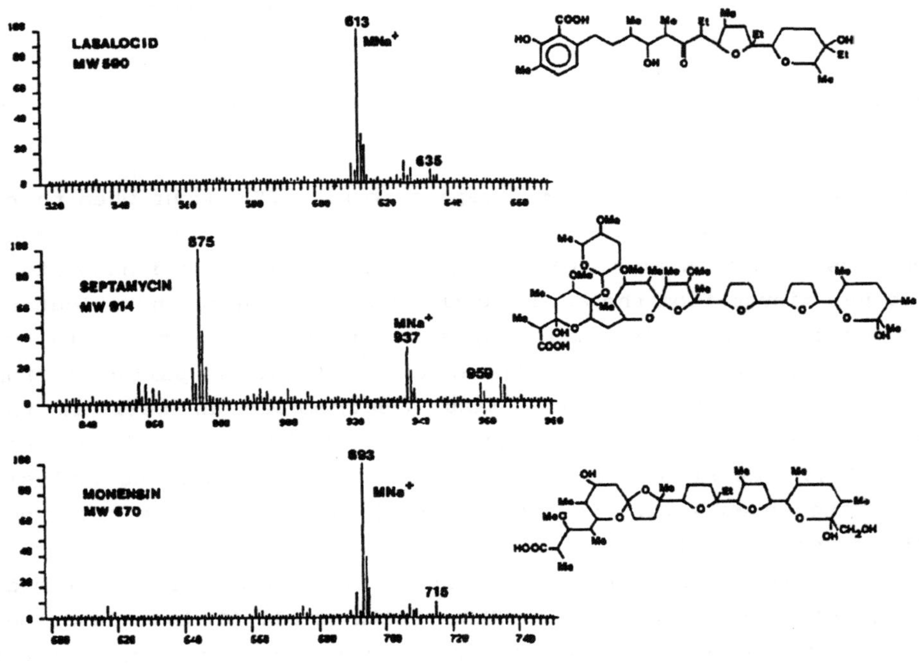

Figure 13. FAB mass spectra of compounds separated by TLC and then transfer of the solid support coating to a FAB probe tip. (Reproduced with permission from reference 12).

Shizukuishi et al. (5) have developed an integrated and automated system for TLC/MS. This technique utilizes a SIMS source and a mechanical system to move the TLC plate in the ion source across the ion beam. This is shown schematically in Figure 14. An example of the mass spectra produced is shown in Figure 15 for the analysis of phospholipids analyzed in the negative ion mode. Again intense molecular species were recorded with some fragmentation and low

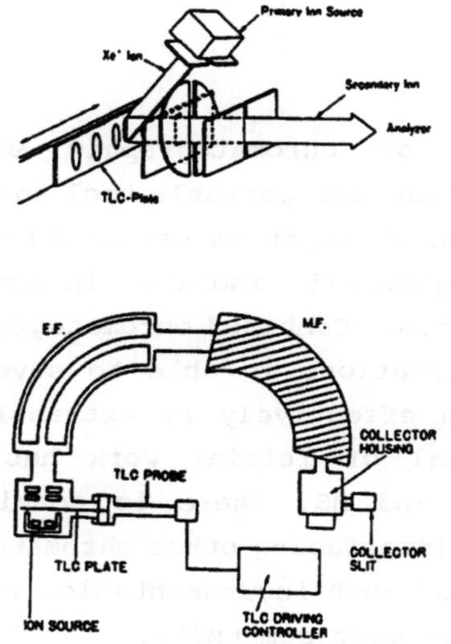

Figure 14. Combined TLC/MS analysis system showing (top) the source region and (bottom) overall instrument diagram. (Reproduced with permission from reference 13).

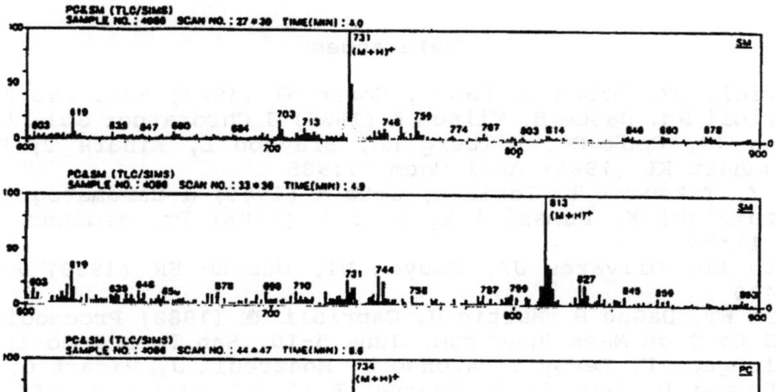

Figure 15. Mass spectra obtained with the TLC/MS system of (top) a sphingomyelin with a C-17 acyl chain; (middle) a sphingomyelin with a C-23 acyl chain; (bottom) a phosphatidylcholine. (Reproduced with permission from reference 13).

levels of background. The approach is unique and quite useful and further work is underway to enhance the sensitivity of the overall procedure.

V. CONCLUSION

The combination of chromatographic separations and mass spectrometry provides an indispensable tool to the biochemist for the structural elucidation of compounds isolated from endogenous sources. These compounds are generally isolated in aqueous systems and are part of complex mixtures. Combined chromatography/mass spectrometry used in on-line applications is able to save enormous amounts of time and can be used effectively at extremely low sample levels. Although a great deal of initial work has been done with the combination of HPLC and MS, there is considerable interest and increased efforts in interfacing other chromatographic procedures as well. It is clear that such instrumentation is needed and will find wide acceptance in the user community.

Acknowledgments

The author wishes to thank Beverly DaGue, Millie Martin, and William Moore for their help in obtaining some of the data presented.

References

1 Caprioli RM, DaGue B, Fan T, Moore WT (1987) BBRC 146:291
2 Caprioli RM, DaGue B, Wilson K (1988) J Chromatogr Sci (in press)
3 Stroh JG, Cook CC, Milberg RM, Brayton L, Kihara T, Huang Z, Rhinehart KL (1985) Anal Chem 57:985
4 Ito Y, Takeuchi T, Ishii D, Goto M (1985) J Chromatogr 346:161
5 Shizukuishi K, Numaziri Y, Kato Y (1986) Proceedings Symp Med Res 11:85
6 Smith RD, Olivares JA, Nguyen NT, Udseth HR (1988) Anal Chem 60:436
7 Moore WT, DaGue B, Martin M, Caprioli RM (1988) Proceedings 36th ASMS Conf on Mass Spectrom, June 5-10, San Francisco 1081
8 Boulenguer P, Leroy Y, Alonso J, Montreuil J, Ricart G, Colbert C, Duquet D, Dewaele C, Fournet B (1988) Anal Bchm 168:164
9 Lisek CA, Bailey J, Aimone LD, Benson L, Yaksh TL, Jardine I (1988) Proceedings 36th ASMS Conf on Mass Spectrom, June 5-10, San Francisco 1083
10 Minard RD, Chin-Fat D, Curry P, Ewing AG (1988) Proceedings 36th ASMS Conf on Mass Spectrom, June 5-10, San Francisco 950
11 Nelson R, Vourous PV, Karger BL, unpublished data
12 Chang TT, Lay JO, Francel RJ (1984) Anal Chem 56:109

The Application of High Mass FAB Mass Spectrometry to Molecular Biology

by

Howard R. Morris, Anne Dell, Maria Panico, Roy McDowell
and Ashraf Chatterjee

Department of Biochemistry, Imperial College of Science and Technology, London SW7 2AZ, U.K.

Introduction

Modern research in molecular biology is devoted to the study of the structures of biopolymers and their interactions with each other and with low molecular weight substrates - the phenomenon of Molecular Recognition - the processes which give functionality to living systems. A deep understanding of the functions of these macromolecules must depend upon understanding the chemistry of these interactions, which itself demands definition of their structures in three dimensions, the fundamental pre-requisite being the basic primary or covalent stucture of the molecule. Mass spectrometry has played an important role in biopolymer analysis for some twenty years now, in particular in protein sequence analysis where it is complementary to classical gas phase Edman sequencing, but also in the analysis of post-translational modifications where classical methods cannot give definitive data. The necessity for accurate protein sequence analysis in molecular biology has never been greater, providing a stark contrast to predictions of obsolescence in the

U. P. Schlunegger (Ed.)
Biologically Active Molecules

early 1970s when the power of the nucleotide sequencing methods was recognised. The fact is that even when the human genome sequencing project gets under way, the nucleotide sequence data will be of limited use without data on the nature of the actual proteins expressed, whether they are glycosylated, proteolytically processed, covalently modified, disulphide-bridged and so on; in other words the vital data on which functionality depends will often not be revealed by a nucleotide sequence. Mass spectrometry has much to offer in this field of research, and early examples of the development and application of new methods from this laboratory include the discovery of γ-carboxylation of glutamic acid, a functional pre-requisite in certain blood coagulation zymogens (1,2), the discovery of 1-3 linked galactosyl N-acetyl galactosamine in plasma glycopeptides (3) and the characterisation of conjugate structures of great biological significance, such as the peptidolipid leukotriene LTD-4 (4) associated with a variety of allergen-induced hypersensitivity reactions such as asthma in man.

Apart from the ability to study non-protein modifications (5), e.g. sulphation, phosphorylation or glycosylation, the other strength of a mass spectrometric approach is the ability to analyse mixtures, something which cannot normally be achieved by classical methods. Since purification is often the rate determining step in structural analysis (and causes decreasing yields), a technique capable of analysing mixtures has obvious applications. Direct mixture analysis by mass spectrometry was developed by us in the mid 1960s for the study of protein digests (6). There is only a minimal need for purification: several peptides can be analysed as the acetyl permethyl derivatives, and usually completely sequenced without purifying individual components in a mixture. Mixture analysis strategies have been used over two decades to solve structures ranging from silk fibroin (6) to ribitol dehydrogenase (7), chloramphenicol transacetylase (8) and the enkephalins (9). These studies were, however, technically very difficult, and consumed more sample than is currently necessary.

More recently, two advances have combined to allow more widespread application of mass spectrometric methods to biopolymer problems. These are the development of High Field (10,11) and other high-mass instruments (12), and the introduction of Fast Atom Bombardment (13-15) as a new ionization method. Using these techniques, it is now possible to exploit the old mixture analysis strategies in protein, glycoprotein and other post-translational analysis at the picomole-nanomole level (16-18).

Importantly, the use of high mass fast atom bombardment mass spectrometry in conjunction with classical protein and carbohydrate chemistry has proved to be a powerful and cost effective method for structural analysis of genetically engineered proteins and glycoproteins of interest to the biotechnology industry (19,20). In relation to this, we have been actively developing specific methods for determining the identity of blocked N-terminal residues and for detecting the presence or absence of C-terminal fragments or "ragged ends" - two areas that pose extreme problems for conventional classical techniques. In addition, characterisation of recombinant proteins for both research and regulatory body approval will require determination of post-translational modification such as glycosylation or S-S bridge assignment. We have developed new approaches to solving these problems, based on FABMS mixture analysis procedures, christened FAB mapping (14,20,21) and the novel strategies, described below, constitute a powerful new tool for protein and glycoprotein engineers and chemists.

FAB Mapping

An important application of the general FAB mapping strategy is in the process or quality control analysis of recombinant genetically engineered proteins. The majority of the structure can be screened by combining data from two or more simple FAB maps after, for example, tryptic, cyanogen bromide, chymotryptic or V8 protease digestion to generate peptide mixtures. Full sequence confirmation on a molecule of the - Interferon type (18,000 Da) is normally provided from the above analysis, sometimes incorporating partial purification (HPLC) of digests, in order to minimise the partial suppression phenomenon described in our early FAB studies (14). The procedure thus complements simple HPLC tryptic profiles by giving an added dimension of definitive mass assignment in quality control and product verification. The same protocol is also a research tool at the pre-production stage, and further examples of the power of the FAB mapping method are described below.

Figure 1 shows the first biotechnology application of FAB mapping to the analysis of a fusion product of urogastrone (epidermal cell growth factor) in 1982 (22). Our approach was to treat the de-salted protein sample, following reduction and carboxymethylation of cysteine residues, with the enzyme trypsin, which cleaves only on the C-terminal side of Lysine or Arginine. The intact digest was mapped by FABMS without further fractionation and the spectrum shows the molecular weight signals [M+H] of all the tryptic peptides derived from the sample. These can be mapped onto the anticipated structure, demonstrating that the primary sequence is

Figure 1:

FAB MAP OF RECOMBINANT EPIDERMAL GROWTH FACTOR

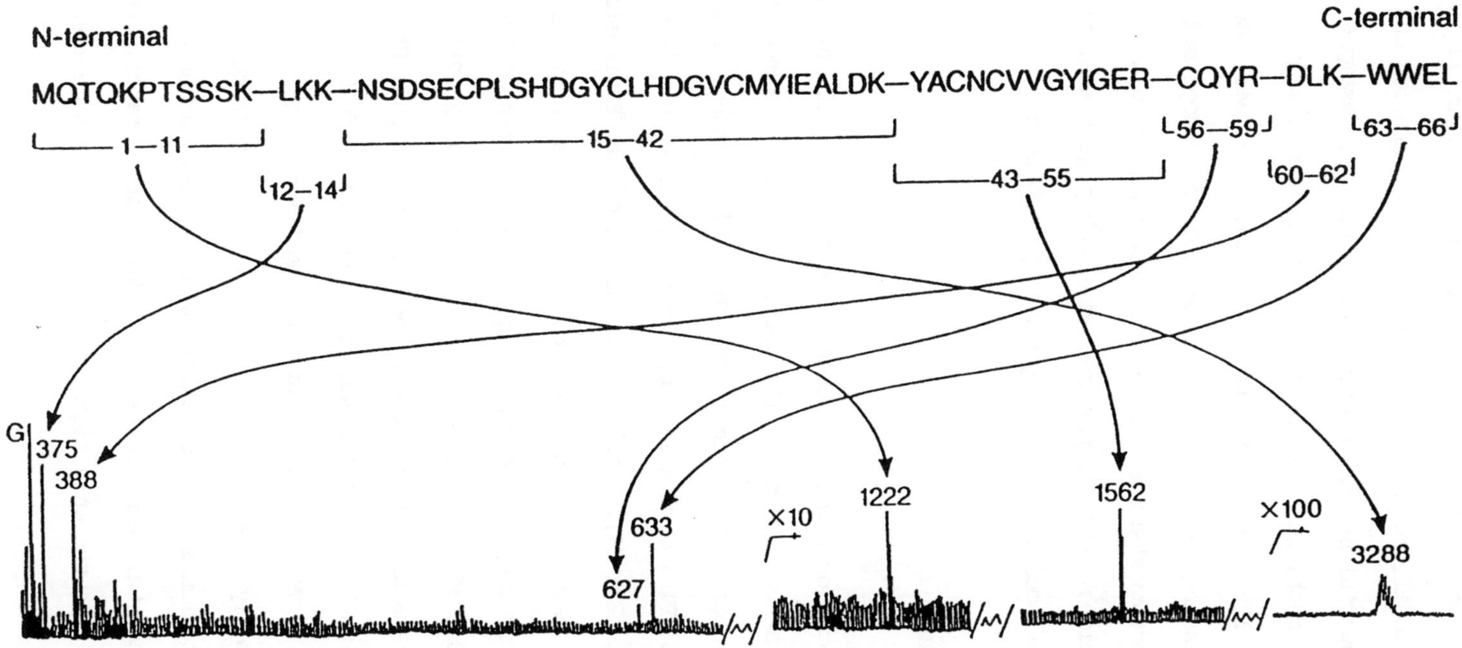

Fast atom bombardment mapping strategy. The protein of interest is cleaved using specific enzymic or chemical methods. The peptide fragments are analysed by direct FAB-MS. The peaks in the mass/charge (m/z) spectrum can be assigned to particular peptides. In this example, a carboxymethylated urogastrone fusion product was digested with trypsin which cleaves after lysine and arginine residues. Subsequent digestion of the tryptic fragments with carboxypeptidase B can often be used to identifiy the C-terminal peptide; unless the C-terminal amino acid is lysine or arginine, the C-terminal peptide will be unaltered in mass by carboxypeptidase B digestion. All the other fragments (unless possessing penultimate proline) will have their C-terminal lysine or arginine residues (highlighted) removed by this treatment.

correct. If unexpected signals are observed, errors of translation, deletion, insertion, point mutation (excepting inversions in sequence), post-translational modification or processing, can be detected and assigned by this powerful new FAB Mapping technology. For example, suppose that the N-terminal methionine had been formylated, then a signal 28 mass units higher at m/z 1250 rather than m/z 1222 would have been observed. Similarly, if processing had taken place this could be recognised by a shift in the mass observed either to lower masses than calculated, for loss of signal peptides, or to higher mass due to a post-translational modification such as addition of carbohydrate, phosphate etc.

C-Terminal Sequencing

Biochemical, enzymatic or chemical procedures such as tryptic or cyanogen bromide digestion can be used to verify assignments in a FAB map. Mass shifts after Edman degradation (observing the peptides rather than the phenylthiohydantoin (PTH) derivatives of the amino acids) can lead to sequence assignment (23, 24). Similarly, carboxypeptidase B digestion of the tryptic digest (carboxypeptidase B removes C-terminal arginine or lysine residues) produces, in the resulting FAB map, shifts of 128 or 156 mass units - for loss of lysine or arginine respectively (20,21). We have demonstrated that this novel procedure of producing a tryptic FAB map of a protein and comparing that with the map produced after a further step of carboxypeptidase B is very useful in identifying the true carboxy terminus of a protein - normally a very difficult task : the peptide corresponding to the original C-terminus will not shift in mass after carboxypeptidase B digestion (20) unless the terminus is Lys or Arg. For instance, the signal at m/z 633 in Fig. 1 is the only signal to remain at the same mass after subdigestion, and corresponds to the carboxy terminal peptide (residues 63-66). The procedure can be modified with other enzyme combinations if the C-terminus is anticipated to be Arg or Lys, and importantly only a few tens of picomoles of sample is consumed in this analysis.

This powerful procedure has now been applied to some of the most daunting protein chemical problems, such as the search for partial amino acid sequences in human low density lipoprotein APO B (molecular weight c. 500 kD) used to find the gene coding for the full protein sequence (25,26). In a 1983 study to determine peptide sequences in APO B, a peptide isolated by HPLC from a tryptic digest of the molecule retained the same mass after FAB mapping following the carboxypeptidase B subdigest procedure described above. This made the peptide a prime candidate for the true carboxy terminus of the whole 4,500 residue molecule. Sequence ion analysis of the peptide fragmentation pattern in the FAB spectrum produced a sequence Leu-Ala-Pro-(Glu,Gly,Leu)-Thr-Leu-Leu-Leu. The bracketed region came from a combination of amino acid analysis and mass data, the quasi-molecular ion [M+H]+ at 1039 calculating for Glu rather than Gln in the sequence. The amino acid analysis also showed a Leu : Ile ratio of 3 : 2 (these amino acids have the same mass and are not distinguished by conventional MS). When the nucleotide sequences of cDNA clones were determined some two years later, the apparent C-terminal residues numbered 4527 to 4536 were translated as Leu-Ala-Pro-Gly-Glu-Leu-Thr-Ile-Ile-Leu, and the MS data that predicted this sequence thus proved invaluable in verifying the C-terminal sequence and the important fact that the protein had not been further processed (25,26).

C-Terminal "Ragged Ends" and Blocked N-Termini

When additional and unexpected signals appear in a FAB map, they can often be interpreted simply and quickly by using a software search of the protein sequence using the peptide mass observed in the spectrum. Sometimes such signals will be due to minor enzymatic cleavages (e.g. chymotryptic activity in the trypsin used) or to impurities, but occasionally these may arise by pre-processing of the sample (naturally or otherwise) to produce truncated species. This can lead to "ragged ends", mixtures of whole and truncated molecules. Ascertaining the existence or absence of truncated species is important in biopolymer characterisation, since, for example, they may not have the correct biological activity or may be antigenic. The analogy in carbohydrate analysis is the presence or absence of sialylation or fucosylation of non-reducing end sugars in glycoproteins. In protein work, if the truncation is N-terminal, and if the N-terminus of the protein is not blocked (with, for instance, an acetyl group), the ragged ends may be assignable by careful examination of gas phase Edman data. This is not usually possible, however, for C-terminal assignment, since it is extremely difficult to know whether a decrease in PTH amino acid yield at a particular step is not just a consequence of the Edman chemistry or the gas phase instrumentation. We have developed a new procedure based on FAB mapping that allows solution of this difficult problem (27).

Because of the suppression phenomenon observed in complex mixtures (14,20), in which not every component in the mixture is necessarily observed, gel filtration or HPLC purification (or partial purification) of a suitable digest is sometimes carried out prior to FABMS analysis to create simpler mixtures. The digest, either chemical (e.g. CNBr) or

enzymatic, is specifically chosen to cut the protein some distance away from the anticipated carboxy terminus. FAB mass spectra of individual fractions of a semi-purified digest will then reveal the integrity of the C-terminus, since the signals observed will fit either for the intact C-terminal sequence and/or for truncated versions of it. Once actual C-terminal species have been identified, it is relatively simple to interpret a suitable carboxypeptidase/amino acid analysis experiment to accurately quantitate the molecular species present.
This work is illustrated for the C-terminal analysis of a recombinant interferon- (Fig.2) (27).

Another important application of FABMS is in the analysis of proteins which possess blocked N-termini of which are blocked, thus preventing sequencing by Edman degradation. Here the definitive nature of mass spectrometry comes into its own : it can detect the presence of a blocking group, and actually allow chemical identification (28). An example of this type of study is shown for the amino-terminal analysis of recombinant α-1-antitrypsin in Fig.3 (29). The expected N terminus of this 394 residue recombinant protein, Glu-Asp-Pro-Gln-Glu-Asp-Ala-Ala-Gln-Lys (m/z 1058), was not mapped in the original FABMS spectra, although over 90% of the molecular structure was confirmed. Because of the partial suppression phenomenon, a search for the amino-terminal peptide by FABMS analysis of an HPLC purified tryptic digest was initiated. This resulted in the discovery of a signal at m/z 1231 in the spectrum of HPLC fraction 21 which could not be mapped onto the anticipated structure using a simple computer program to search for sequences corresponding to the M+H signals in the mass spectrum (Fig 3a). Furthermore, the anticipated N-terminal tryptic peptide (m/z 1058) again could not be found, raising the

Figure 2:

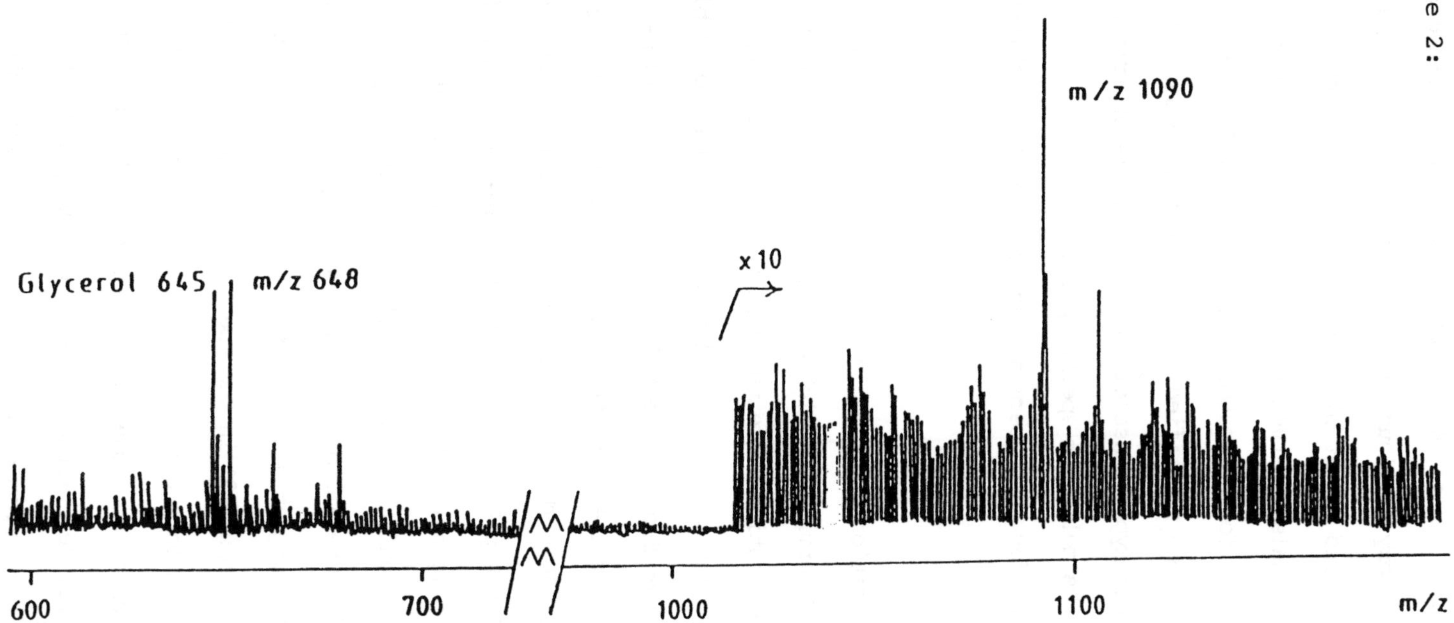

(a) C-terminal analysis of a recombinant interferon-γ. IFN-γ was digested by CNBr and the digest was then passed through a gel column to remove high molecular weight fragments. The 'low mass' (<3 kDa) fractions were then screened by HM-FABMS and fraction 20 (together with adjacent fractions) showed evidence for the 'ragged end' phenomenon. The signals of interest are m/z 648 and m/z 1090. The signal at m/z 1090 is the intact C-terminal signal anticipated for the sequence IFN-γ 138-146. The signal at m/z 648 shows that a proportion of the IFN-γ molecules in the original sample studied were truncated at residue Arg 142 since m/z 648 corresponds to the sequence IFN-γ 138-142, Leu-Phe-Arg-Gly-Arg.

Figure 3:

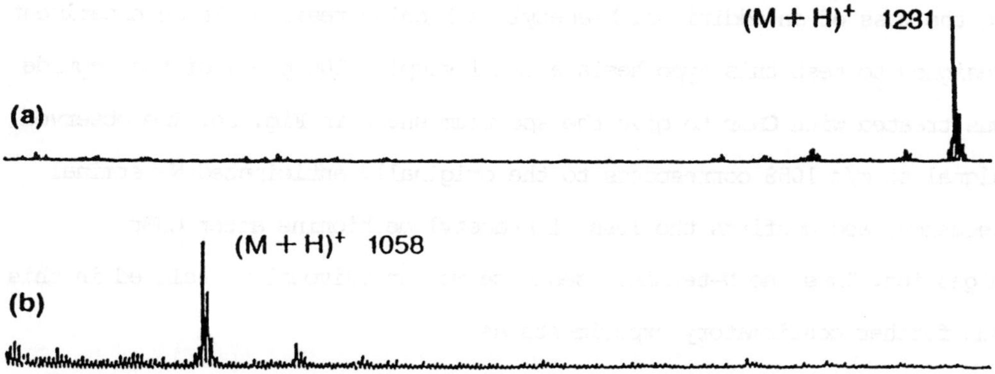

Analysis of blocked N-termini of peptides. (a) The signal at m/z 1231 of a fast atom bombardment mass spectrum of α1-antitrypsin could not be mapped onto the anticipated structure of the protein. The anticipated N-terminal peptide signal was (m/z 1058) absent. (b) Following CNBr treatment, the anticipated signal appeared. The difference in the two masses corresponds to the mass of an N-acetyl methionine residue, thus assigned as the N-terminal amino acid.

possibility that the m/z 1231 signal was a modified version of the N-terminal peptide : the mass difference (m/z 1231 - m/z 1058) corresponds to the mass of an additional N-acetyl methionine residue. In an experiment designed to test this hypothesis a small sample (100 pmol) of the peptide was treated with CNBr to give the spectrum shown in Fig. 3b. The observed signal at m/z 1058 corresponds to the originally anticipated N-terminal sequence, and confirms the loss of N-acetyl methionine after CNBr digestion. Thus the N-terminal sequence was unequivocally assigned in this and further confirmatory experiments as
N-Acetyl-Met-Glu-Asp-Pro-Gln-Gly-Asp-Ala-Ala-Gln-Lys.

Sequencing Unknowns

The necessity to sequence unknown proteins has diminished considerably with the advent of fast and highly sensitive nucleotide sequencing procedures, and certainly few proteins would now need to be completely characterised at the protein sequence level by gas phase sequencing or any other protein method. That is not to say that peptide sequencing has become unimportant, in fact it is essential in many circumstances, but the requirements demanded of it should be understood in order not to waste time on unnecessarily elaborate strategies. In many studies mass alone is sufficient information to check a known sequence such as that of a genetically engineered protein derived for example from nucleotide sequence data. Small stretches of protein sequence are sometimes essential however as the basis for construction of nucleotide probes to allow isolation of the gene coding for a particular protein, and the low density Apolipoprotein B example above was a particularly exciting case of the search for a gene coding for a medically important protein of unknown sequence. In our first studies of the FAB-MS of peptides we described the interesting phenomenon that whilst peptides in mixtures do not give extensive fragment ion information at the 500 picomole level, if the individual components are isolated by HPLC, then fragment ions can be observed in the normal spectrum, usually sufficient to determine 6 to 12 residues of sequence. This natural fragmentation is present in all peptides of 3000 daltons or less provided they are sufficiently pure, and an important point is that 6 to 8 residues of sequence is all that is required to allow construction of a nucleotide probe in many circumstances. In other types of problem, for example the discovery of new neuropeptide or hormone factors, the full determination of amino acid sequence is often the most efficient solution of the structural problem. The approach to such problems, varies depending upon molecular weight, but if a mass spectrometric solution is sought, perhaps because of anticipated post-translational modifications, then the general solution will require determination of the molecular weight and partial sequence from fragment ions, followed by enzymatic or chemical digestion to produce overlaps to complete the sequence. Alternatively one could utilise MS-MS or the new Linked Scan procedures (see later) to enhance the fragmentation observed.

The first neuropeptide of completely unknown structure to be characterised by FAB-MS was one isolated in 1983 from the giant neurone of the gastropod Aplysia (30). In this study we demonstrated the viability of the strategy described above, in exploiting the natural fragmentation sequence ions and completing the sequence using data derived from simple enzymatic cleavages. We had previously determined the fragmentation pathways of peptides ionised by FAB, and defined them as chemical ionisation pathways (14), best rationalised as resulting from protonation

Figure 4:

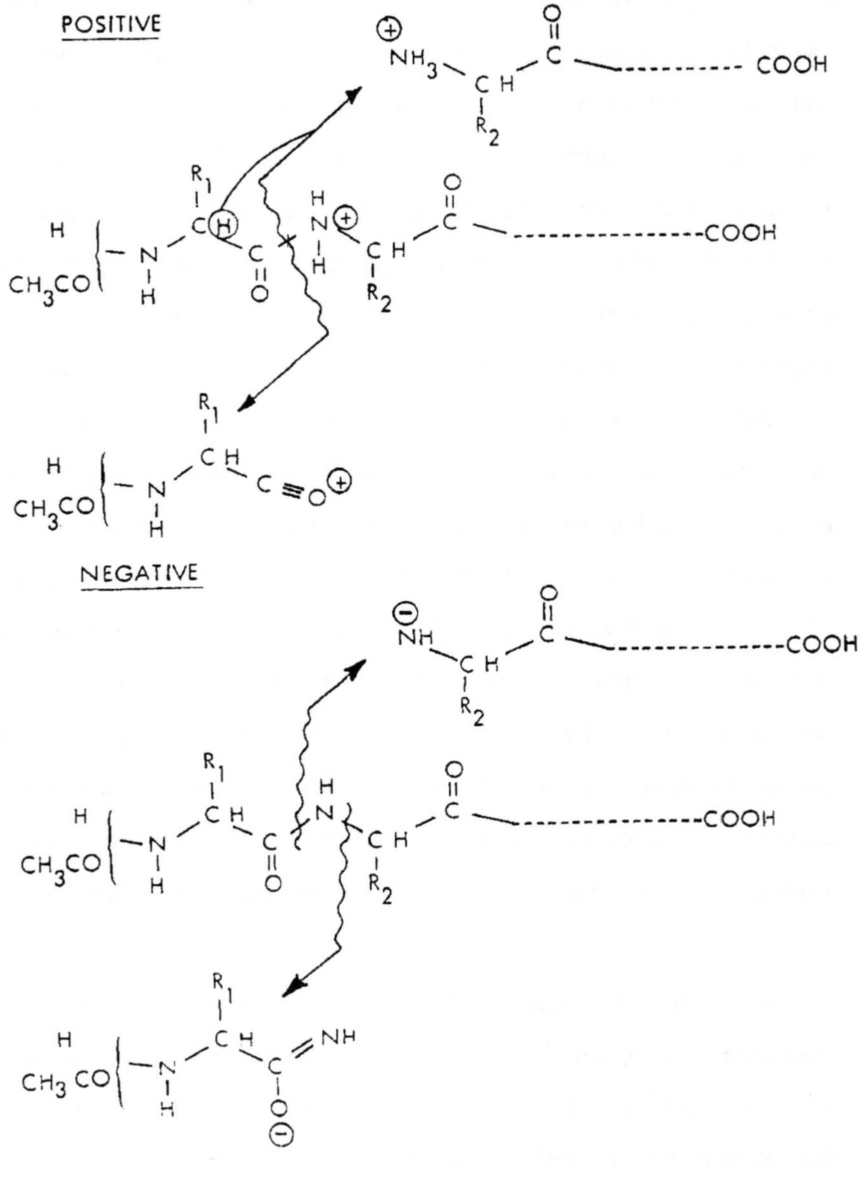

Proposed mechanism for fragmentation of peptides under fast atom bombardment conditions, in both the positive and negative ion modes. The fragment ions observed can be rationalised via the chemical ionisation mechanisms suggested.

on the amide nitrogen of the peptide bond leading to the production of N- and C- terminal fragments as shown in Figure 4. In the Aplysia study, these were differentiated by first treating the sample with a 1:1 mixture of acetic anhydride:d-6 acetic anhydride in methanol to label any free amino group at the N-terminus. The spectrum showed the N-terminal fragment ions as 1:1 doublets 3 amu apart and the C-terminal sequence ions as singlets (C-terminal ammonium ions) with accompanying alkyl satellites 15 amu lower in mass. From these and a simple tryptic fragment molecular ion, a unique sequence could be assigned to this new peptide as the C-terminally amidated structure :
Met-Asn-Tyr-Leu-Ala-Phe-Pro-Arg-Met-amide.(30)

In a similar study, FAB mapping, combined with simple enzymatic digestion and the analysis of natural fragmentation led to the structure determination of Human Calcitonin Gene-Related Peptide (31) as a 37 residue C-terminally amidated structure with an S-S bridge between residues 2 and 7 (Figure 5). It was interesting to observe in this study that normal fragment ions (not collisionally induced) are present in the mass spectrum of the intact 37 residue peptide giving 12 residues of sequence prior to further degradation work.

In some studies, partial sequence is required to determine the position of covalent modification in a protein, which may have been carried out to demonstrate <u>cross-linking</u> to another biopolymer or to establish the <u>active site</u> of an enzyme. An example of the latter comes from our work on the enzyme Chloramphenicol Transacetylase (CAT) whose basic protein sequence we determined in the 1970s in a combined mixture analysis mass spectrometry/nucleotide sequence approach (8). The crystal structure was

Figure 5:

ALA-CYS-ASP-THR-ALA-THR-CYS-VAL-THR-HIS-

ARG-LEU-ALA-GLY-LEU-LEU-SER-ARG-SER-GLY-

GLY-VAL-VAL-LYS-ASN-ASN-PHE-VAL-PRO-THR-

ASN-VAL-GLY-SER-LYS-ALA-PHE-amide

Sequence of Calcitonin Gene-Related Peptide derived from human medullary thyroid carcinoma tissue.

attempted on a more easily crystallisable CAT from another strain of organism. In order to assist interpretation of the electron density maps, it was important to locate the active site for binding the substrate chloramphenicol to the enzyme. A substrate analogue approach was used (32) in which a bromomethyl ketone group was used to induce nucleophilic attack by the enzyme on the substrate once positioned in the active site pocket to produce an irreversible site specific modification. The approach we used to locate which amino acid residue in the enzyme sequence had attacked the substrate was one of proteolytic digestion of the modified CAT, partial fractionation on HPLC followed by an initial molecular weight screening of HPLC fractions by FAB to locate signals whose masses did not map onto the known CAT protein structure. Several such signals were found in different fractions, indicative not of different binding sites of the substrate analogue but of partial tryptic digestion of the undenatured CAT enzyme. One such fraction contained, amongst other signals corresponding to expected peptide molecular weights, a signal at m/z 1877 (Fig 6) which did not fit the CAT structure on a simple computer search but was 58 amu higher in mass than one of the expected tryptic peptides in the sequence (Fig 7). A 58 dalton addition is the expected modification of the active site, since once bound to the enzyme, the chloramphenicol analogue is hydrolysed to leave a carboxymethyl tag as the site label. The sequence corresponding to m/z 1877 contains three potential nucleophiles, all Histidine residues, which may have attacked the substrate, the question being which one carried the carboxymethyl covalent tag? This question was resolved quite simply by carrying out several steps of Edman degradation on the peptide followed by re-examination of the remaining peptide molecular weight by FAB. After 5 steps of Edman the m/z value for the remaining peptide showed that the first Histidine in the sequence had a

Figure 6:

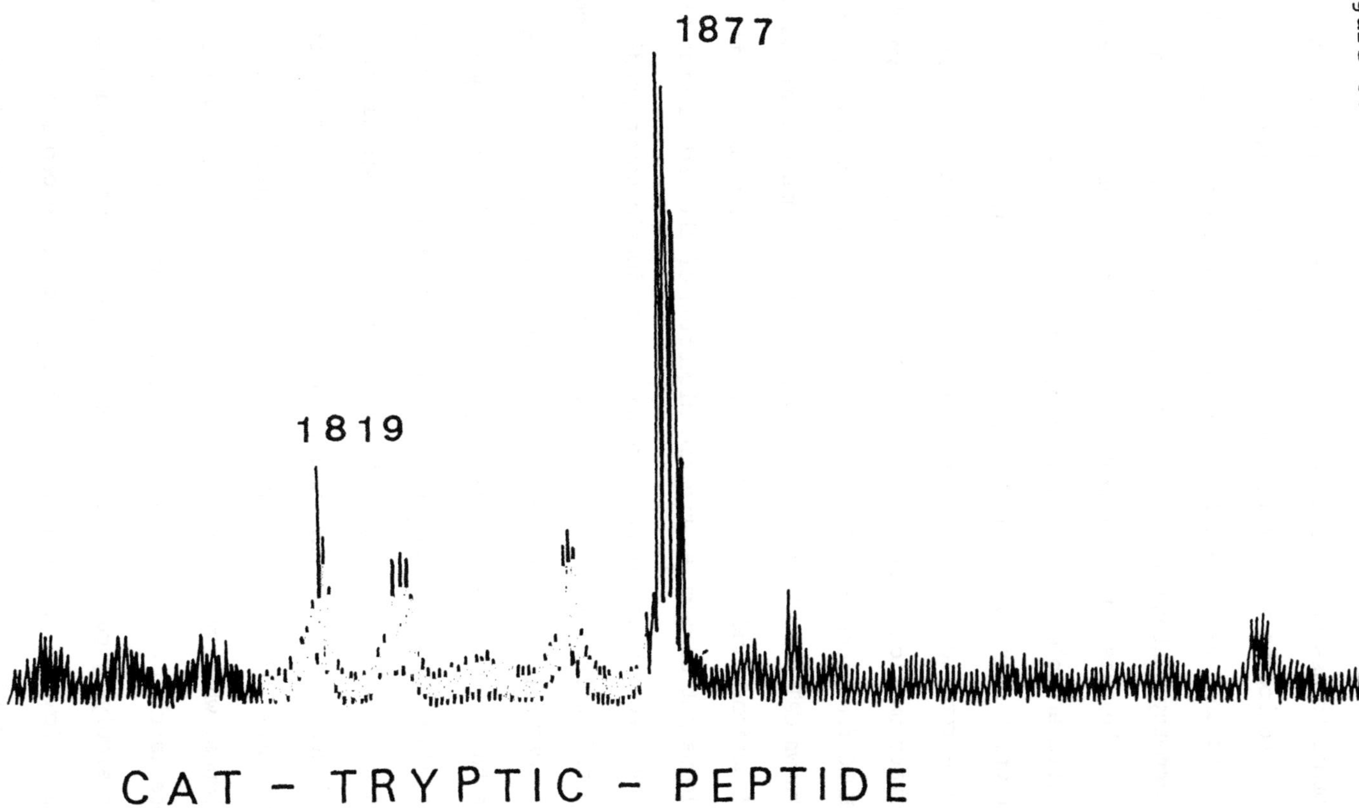

Partial FAB mass spectrum of an HPLC fraction of a tryptic digest of chloramphenicol transacetylase (CAT) which had been pretreated with the substrate analogue and treated with mild base to release a carboxamethyl adduct. Several other peptide M+H signals are present elsewhere in the same spectrum, but these do not affect the subsequent protocol on the mixture.

Figure 7:

SUMMARY AND INTERPRETATION OF EDMAN DATA

```
                ▲?  ▲?          ▲               ▲?
Ser-Val-Gln-Val-His-His-Ala-Val-Cys-Asp-Gly-Phe-His-Val-Ala-Arg.   (M+H)+ I877

[minus Ser-Val] Gln-Val-His-His-Ala-Val-Cys-Asp-Gly-Phe-His-Val-Ala-Arg.   (M+H)+ I69I

[minus Gln-Val-His]  His-Ala-Val-Cys-Asp-Gly-Phe-His-Val-Ala-Arg.   (M+H)+ I327

[minus CMHis]  Ala-Val-Cys-Asp-Gly-Phe-His-Val-Ala-Arg.   (M+H)+ II32
```

Part of the CAT sequence which corresponds to a computer search for M+H 1877 incorporating <u>one</u> carboxymethyl group somewhere in the sequence ▲. Steps of Edman Degradation followed by further FAB analysis were used to locate the covalent modification and thus the active site residue of the enzyme.

normal mass of 137. However, after the sixth step of Edman, the signal shifted from m/z 1327 to m/z 1132 , a loss of 195 amu due to a carboxymethyl Histidine residue (Fig 7). It can thus be concluded, quite definitively, that Histidine-189 in the sequence is the active site residue in Chloramphenicol Transacetylase (Fig 8), and this has since been confirmed in the X-ray diffraction study of the crystal structure.

Biomedical Research

Many of the studies described in this article could be listed under a general heading of biomedical analysis, the low density lipoprotein LDL involved in plasma cholesterol transport and the characterisation of the putative asthma factor, Leukotriene D-4, being obvious examples. Some of the most exciting and worthwhile research involving mass spectrometry is in areas in which we can attempt to give a chemical and molecular description of, and explanation for, disease states. The prospect then exists for intervention in a rational manner either to replace defective templates (gene therapy) or end-product molecules (drug therapy including recombinants), or to target enzyme deficiencies in a biosynthetic pathway. An example of one such study concerns a distressing inherited abnormality in the blood coagulation control system in which even young family members have a predisposition towards thromboses and have reduced life expectancy especially if the condition is not diagnosed and intensive drug therapy not provided. The blood coagulation system - leading ultimately to the action of thrombin, a large glycoprotein, on fibrinogen to produce cross-linked fibrin peptides forming the framework for the clot formation at the site of a wound - is an extremely complex array of glycoprotein factors which require activation at each step of the process of clot

Figure 8:

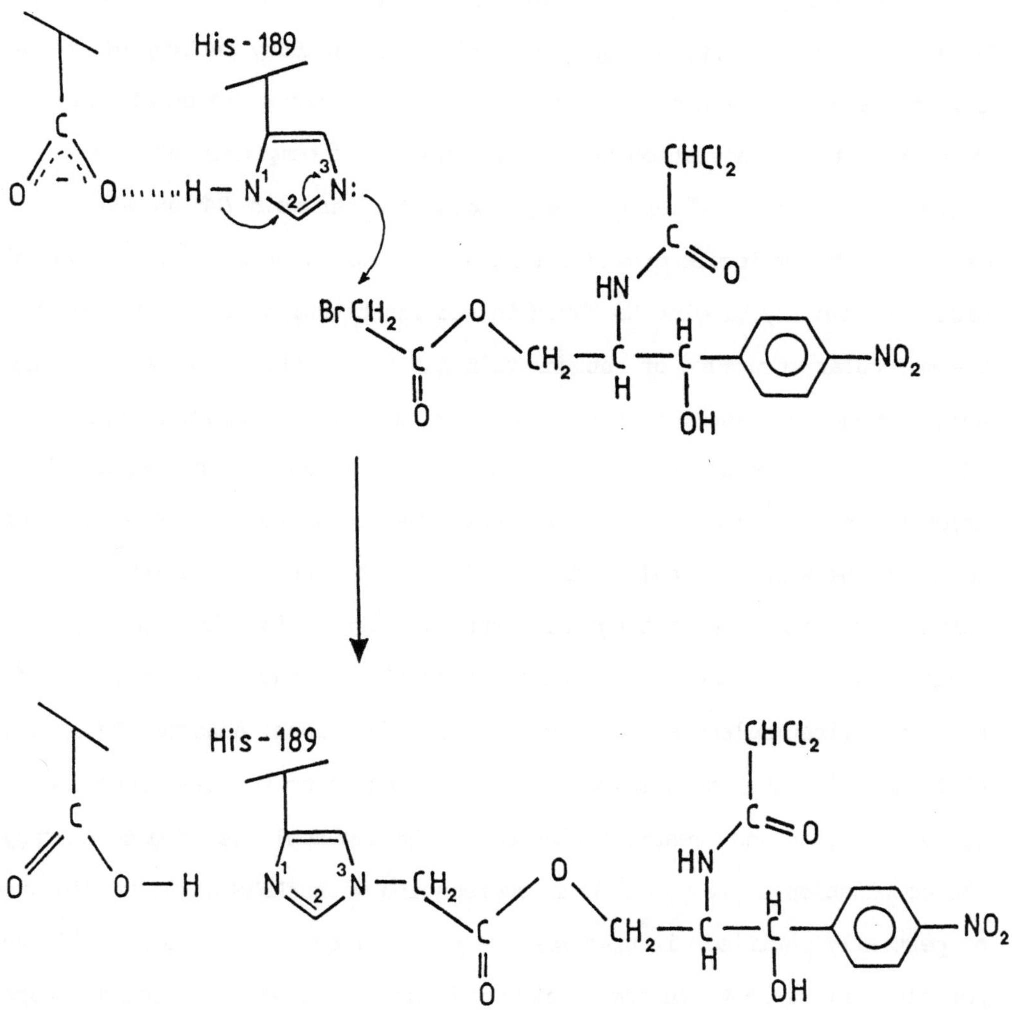

HIS-189 was identified at the active site residue of CAT. The mechanism by which the substrate analogue labelling of the enzyme takes place is illustrated. A subsequent step of mild base hydrolysis leaves a carboxymethyl residue at the enzyme active site.

formation. There is thus an internal control to minimise the possibility of a potentially fatal thrombus formation in the absence of an open wound. More recently, however, an equally intricate inhibitory pathway of complex glycoproteins has been discovered which gives an extra degree of control over the active coagulation zymogens, and a major component of this inhibitory system is a large glycoprotein, Antithrombin (MW 65,000 daltons). This molecule provides a final line of defence against unwanted clot formation by binding to thrombin and inactivating it. A study of the the molecular structure of antithrombin has now identified a set of amino acid sequence variants, arising from point mutational errors in the coding sequence of the gene, in the antithrombin from those patients with this inherited abnormality. Importantly, these variants have all been found to occur at the same critical binding region in the thrombin/antithrombin complex, reducing the binding efficiency and preventing the proteolytic event known to be associated with antithrombin binding. This is believed to be the first molecular description for a functional abnormality leading to thrombosis, and the possibilities are encouraging for new probes for early detection and treatment, including the eventual use of gene therapy. The combination of steps of Edman degradation with FABMS of unfractionated or partially purified mixtures was the powerful one chosen to identify the pathological variants in the antithrombin study (33,34). Peptide FAB maps were created from CNBr, tryptic and V8 protease digests, and potential variant amino acid sequences located by a sequence search computer program. In an example of one particular study, the data produced (Fig 9) allowed the location of two important Antithrombin variants to residues 371 to 399 in the sequence. Using the subtractive Edman strategy described for the CAT work above, 11 steps of Edman followed by MS conclusively proved that the variant structures contained respectively a CYS-393 and a

Figure 9:

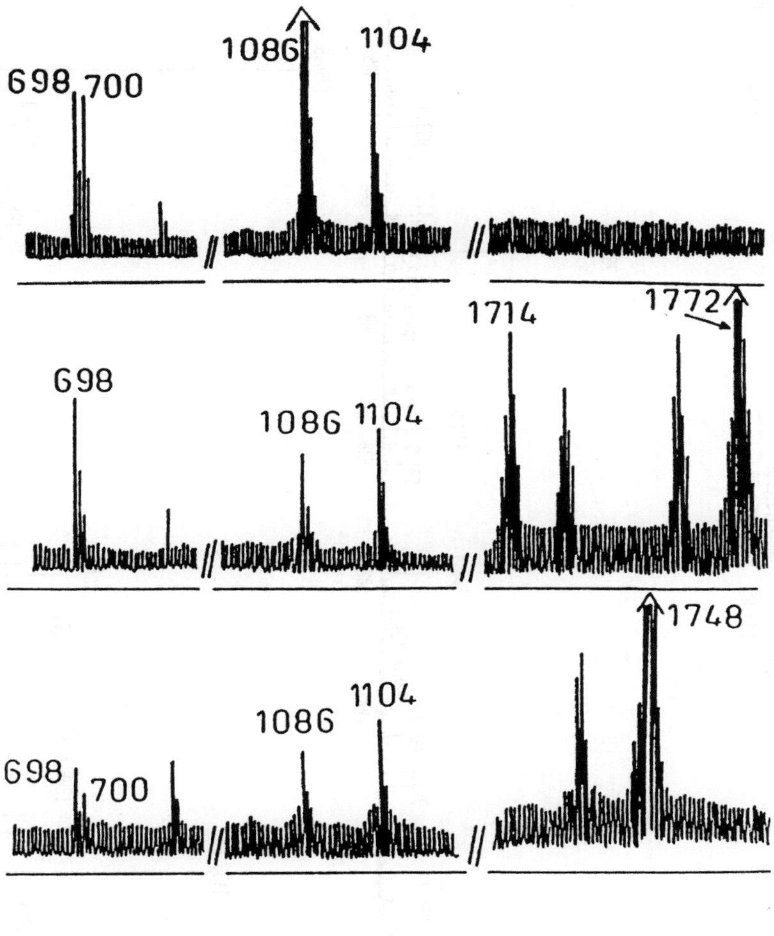

Comparison of the FAB mass spectrum of the V8 protease sub-digests of the tryptic peptides of a CNBr pool from normal antithrombin (upper), antithrombin Northwick Park (middle) and antithrombin Glasgow (lower). Only regions of the spectrum where differences in the digests were observed are illustrated. Unlabelled signals are anticipated satellites of major signals, e.g., -15, -44, +22 amu.

Figure 10:

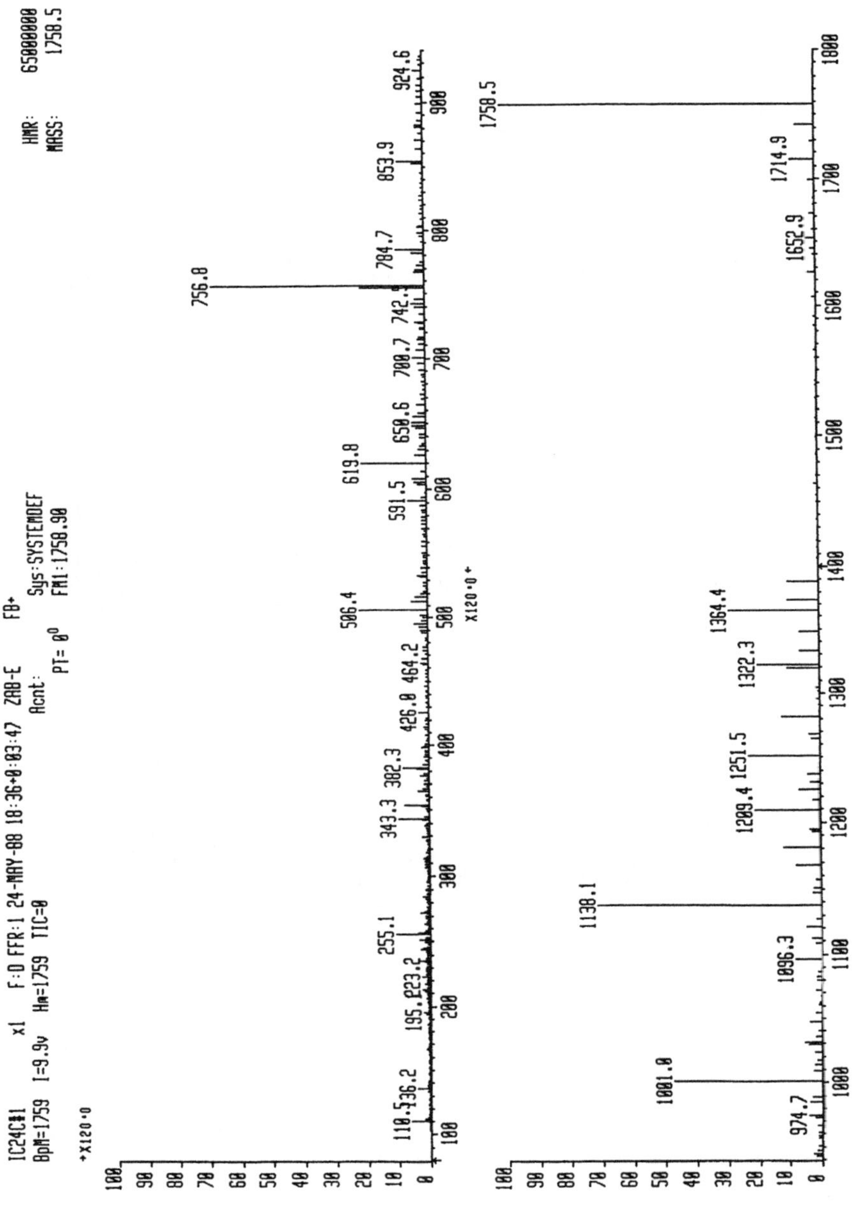

Linked scan mass spectrum taken on the quasimolecular ion of 200 picomoles of Renin Substrate on a ZAB2SE, showing fragment ions from which sequence can be confirmed. Nominal fragment ion masses are given, and the spectrum is the result of six accumulations from one loading of sample.

HIS-393, where the normal structure has ARG-393. This result at the same time gave us an unusually clear insight into why the variation is pathological, since the ARG-393 SER-394 bond is the one split during thrombin binding in the normal physiological situation.

MS/MS and Sequence Determination by Linked-Scan in Mixtures

In 1981, during our initial work examining the advantages and limitations of FAB-MS (14) we carried out some linked scan experiments in an effort to enhance the fragmention data seen in FAB spectra. The results at that time were not encouraging, but we have recently re-evaluated linked scan following sensitivity improvements in the mass spectrometer, and software improvements for the linked scan itself. The results are now very encouraging insofar as we have been able to demonstrate very similar data to that obtained on tandem MS-MS instruments when examining individual peptides by helium collisional activation. More importantly, we have applied the procedure to total enzyme digest mixtures to give enhanced fragmentation on unseparated components of those mixtures. The helium collision spectrum of Renin Substrate (200 picomoles) on the ZAB 2SE is shown in Fig 10 which is the linked scan spectrum on m/z 1758 (M+H). Sequence ions are very prominent, and the spectrum is very comparable with the tandem MS-MS data on the same peptide. Linked scan data on isolated individual peptides can now be valuable therefore in enhancing normal fragmentation, but its real utility may lie in providing collisionally activated spectra from components of mixtures without isolation and without the need for tandem MS-MS instrumentation. Of course, since the window for the linked scan is larger than one can tune to on four sector instrumention the spectrum contains decomposition products from matrix

Figure 11:

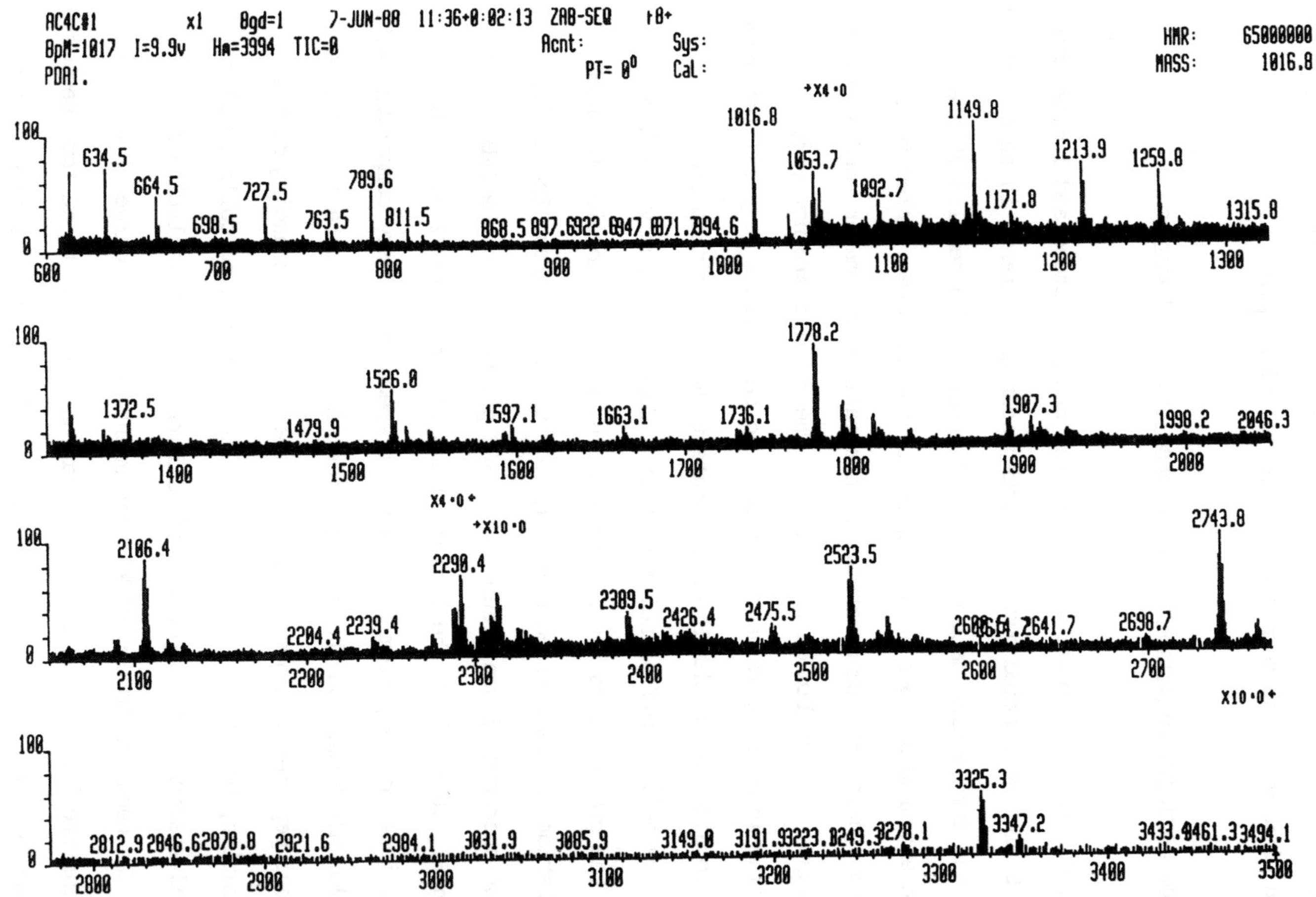

Full FAB map of a tryptic digest of Pyruvate Decarboxylase taken with the high energy cesium gun on a ZAB2SE, prior to linked scan analysis of unknown components in the mixture (Morris H.R., Chatterjee A., Panico M., Green B., Amazonas Almeida de Silva M.A., and Hartley B.S., submitted for publication).

ions, but the data are very usable in most circumstances, as is demonstrated below. In a study on an unfractionated tryptic digest of Pyruvate Decarboxylase we first produced the normal FAB map shown in Figure 11, and screened the M+H signals against the sequence predicted from DNA data. Several of these signals do not match the predicted sequence, and we have studied the linked scan spectra of these signals to test its utility in problem solving. The results are interesting in that each unknown signal was sequenced either by use of linked scan alone or in combination with steps of Edman or CPB on the intact mixture. As an example, the linked scan spectrum of m/z 789 is shown in Figure 12. The sequence LEU-THR-LEU (where LEU may be ILE) is clearly derived from an immediate glance at the strongest signals, and the terminal masses of 259 and 204 amu were then tested to fit a C-terminal LYS or ARG anticipated from a tryptic digest. Only m/z 204 fits for a meaningful sequence, GLY-LYS, thus leading to the sequence ....LEU-THR-LEU-GLY-LYS. At this stage other ions in the spectrum can be used to complete the sequence but it is more efficient to re-screen the overall sequence to look for a fit for the pentapeptide deduced. The search finds MET-SER-GLU-ILE-THR-LEU-GLY-LYS, the predicted N-terminus of the protein, but this does not fit for the mass observed at m/z 789. This mass together with the remaining linked scan signals suggests that the N-terminus of the protein is in fact blocked, and the sequence is ACETYL-SER-GLU-ILE-THR-LEU-GLY-LYS. A single step of Edman on the intact mixture followed by FAB mapping then confirmed that m/z 789 did not move, confirming a blocked N-terminus. In summary, good fragment data is obtainable at least up to m/z 2000, although as with tandem MS-MS all the anticipated fragment ions are not necessarily present in every peptide, depending upon structure. Overall our data suggest that linked scan will

Figure 12:

LINKED SCAN OF ENZYME DIGEST MIXTURE (ZAB 2SE)

m/z 789

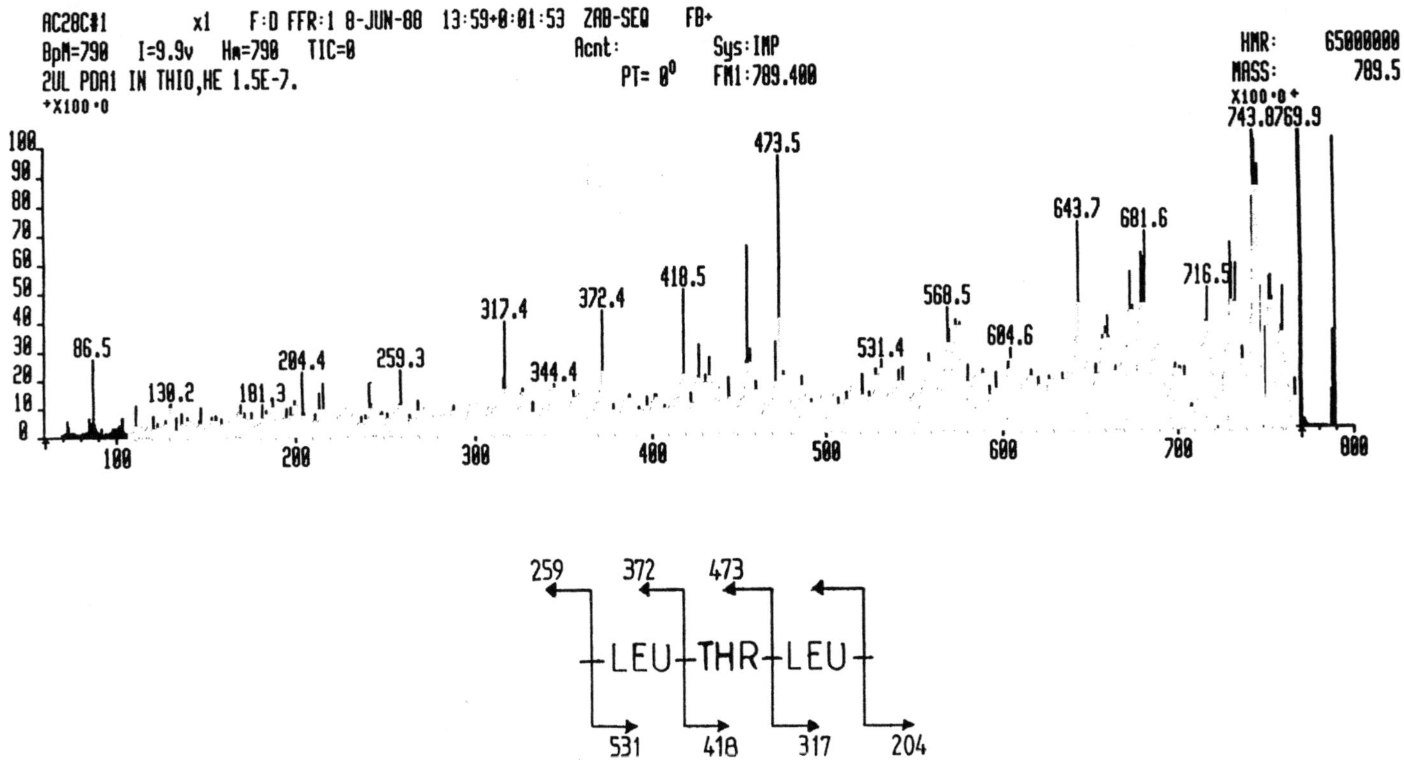

Linked scan spectrum of an unknown component (m/z 789) in the intact tryptic digest of a Pyruvate Decarboxylase, from which the blocked N-terminal amino acid sequence was deduced.

provide an additional useful procedure for the sequence analysis of biopolymers.

Post Translational Modification

Post translational modification of proteins is a very common event in the expression of functional products. In many cases the nascent polypeptide chain undergoes covalent modification such as N-terminal acylation or C-terminal amidation, phosphorylation, glycosylation, carboxylation, S-S bridge formation and/or simple proteolysis to release a targetted product. Examples of post translational modification have already been given under other headings, e.g. N-terminal blocking groups, but here we discuss additional procedures for glycoprotein and S-S bridge analysis
Methods for the mass spectrometric analysis of glycoproteins were first developed some years ago on plasma glycoproteins from Antarctic fish (3). These involved the analysis of acetyl permethyl derivatives of glycopeptides by electron impact and chemical ionisation mass spectrometry both prior to and following release of the O-glycosidically bound carbohydrate from the peptide backbone by beta elimination in base. Similar powerful strategies (Fig 13) can now be used, incorporating high mass FABMS analysis of both free (underivatized) samples,and acetyl or permethyl derivatives in studies to determine sites of glycosylation of both O- and N-linked structures, the identity of terminal non-reducing ends (potentially the most antigenic structures), and the general types of oligosaccharide. Strategies for analysing glycans from glycoproteins can be conveniently divided into those which accomplish an initial screen for presence and type of glycosylation and sites of glycan-protein attachment, and those which, in conjunction with classical carbohydrate procedures,

Figure 13:

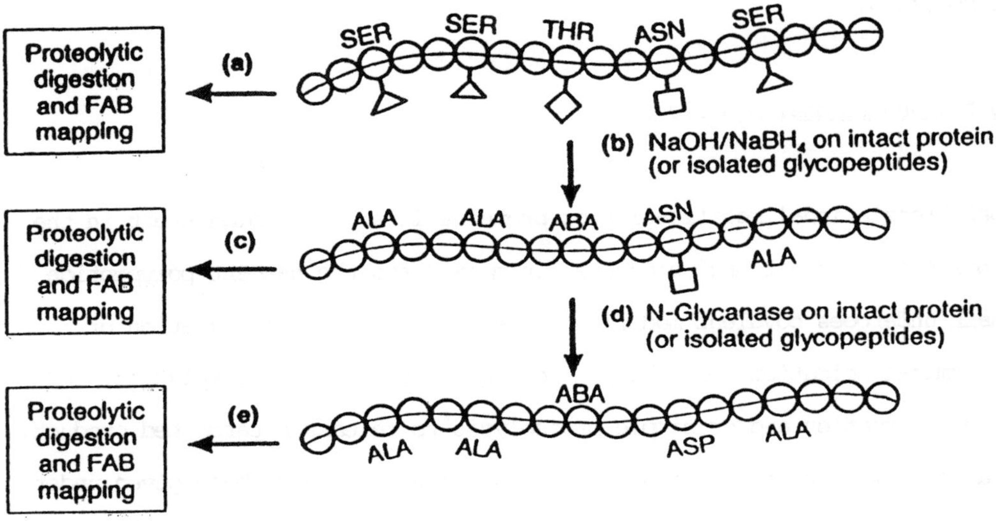

Strategy for the location of O- and N-linked glycosylation sites in a large glycoprotein. Step (a) gives a FAB map baseline. Step (b) removes O-linked sugars, converting Ser to Ala (or ΔAla if base alone is used) and Thr to α-aminobutyric acid, ABA (or ΔABA if base alone is used). The FAB map at step (c) reveals those peptides that were originally O-glycosylated. Removal of the N-linked sugars [step (d)] converts the glycosylated Asn to Asp and the FAB map at step (e) reveals the N-glycosylated peptide.

furnish detailed structural information.

O-linked chains are readily released from the protein backbone using the well established reductive elimination procedure (35). We have exploited this methodology in a FAB based strategy for rapidly screening O-linked glycoproteins to ascertain the compositions of all glycans present, thereby determining the degree of compositional heterogeneity (36).The strategy can also be used to answer the question "is this protein O-glycosylated?". The novelty of our approach lies in the derivatisation step which follows the reductive elimination reaction. The products of the reaction are peracetylated, without prior purification, by the direct addition of the acetylating reagent (an equimolar mixture of trifluoroacetic anhydride and acetic acid) to the dried product mixture. After a short incubation at room temperature the reagents are removed by evaporation and the products partitioned between water and chloroform. The latter contains the peracetylated oligosaccharides free from salts, protein and other impurities and is sufficiently clean for direct FAB analysis. The peracetylated oligosaccharides afford abundant molecular ions from which their compositions in terms of hexoses, aminohexoses etc can be readily computed (37). If information on protein attachment sites, as well as types of glycans, is required the products of reductive elimination can first be separated into a carbohydrate and salt containing fraction and a protein containing fraction using reverse phase chromotography. The carbohydrate fraction is acetylated as described above and the protein is subjected to the FAB mapping procedure described earlier. If necessary, sodium borodeuteride is used in the reduction step in order to facilitate identification of peptides containing reduced side chains, i.e. those which had been glycosylated prior to the reductive elimination. Alternatively, if non-reductive elimination is chosen, the sites of attachment of O-glycans are identified by the presence of dehydro-amino acids in the protein backbone (3).

The simplest procedure for determining whether N-linked glycoproteins contain high mannose, complex, hybrid or lactosaminoglycan structures involves FABMS analyses of aliquots removed from a time course acetolysis of the intact molecule (38). The chloroform soluble fractions will contain peracetylated oligosaccharides cleaved from the glycoprotein by the reagent and their molecular ions will define the type(s) of glycans present. A second approach (36,39) exploits both carbohydrate and protein techniques and is an extension of the FAB mapping strategy described earlier. The glycoprotein is digested with trypsin and a portion of the resulting peptide/glycopeptide mixture is analysed by FAB-MS. The FAB map contains signals from peptide constituents of the mixture. The remainder of the tryptic digest is further digested by N-glycanase and a FAB spectrum of the newly digested sample is acquired. Any new signals appearing after the second digest are derived from peptides originally linked via Asn residues to a carbohydrate chain. These peptides now contain Asp at the original glycosylation site. Signals from released carbohydrates are not seen in the maps since native peptides ionise more readily than native oligosaccharides. The oligosaccharides can, however, be observed by a simple peracetylation reaction followed by chloroform extraction and FABMS of the chloroform soluble material. The molecular and fragment ions present in the FAB spectra will reveal the types of N-glycan present in the glycoprotein.

The most appropriate strategy for analysing O-linked glycans is dependent on the complexity of glycosylation (few or multiple attachment sites) and the extent of knowledge of the protein structure. If the protein sequence is known and a limited number of glycosylation sites are expected, a suitable strategy is one which incorporates both protein and carbohydrate techniques. Glycopeptides are prepared by proteolytic cleavage at sites conveniently close to the glycosylated residues. FABMS of the native glycopeptides affords abundant molecular ions which define the carbohydrate composition and give an indication of heterogeneity. In the first example of the analysis of an unknown potential glycopeptide structure by FABMS, the glycosylation on natural interleukin 2 (IL-2) was discovered (Fig. 14), (40,41). The gas phase sequence of the protein had given a blank in the position 3 PTH in the sequence, indicating the possibility of a post translational modification in this position. FABMS of a peptide isolated by HPLC from a tryptic digest of Interleukin 2 showed an intense quasimolecular ion clearly showing the presence of a glycopeptide and indicating that the carbohydrate moiety was NeuAcHexHexNAc attached to a threonine residue at position 3. Stereochemistry and linkage were later established by classical carbohydrate analysis of the hydrolysed glycopeptide. In further work on natural Interleukins from different cell lines, analysis of the isolated N-terminal tryptic peptide followed by FABMS analysis allowed identification of five variant structures (41), including one containing HexNAc at position 3 in the peptide, one unglycosylated, and two containing truncated peptide sequences.

The beta-cleavage fragment ions (42) observed in the spectra of glycopeptides are indistinguishable from molecular ions of glycopeptides containing truncated carbohydrate chains and to differentiate these possibilities the sample must be purified or derivatised. This is

Figure 14:

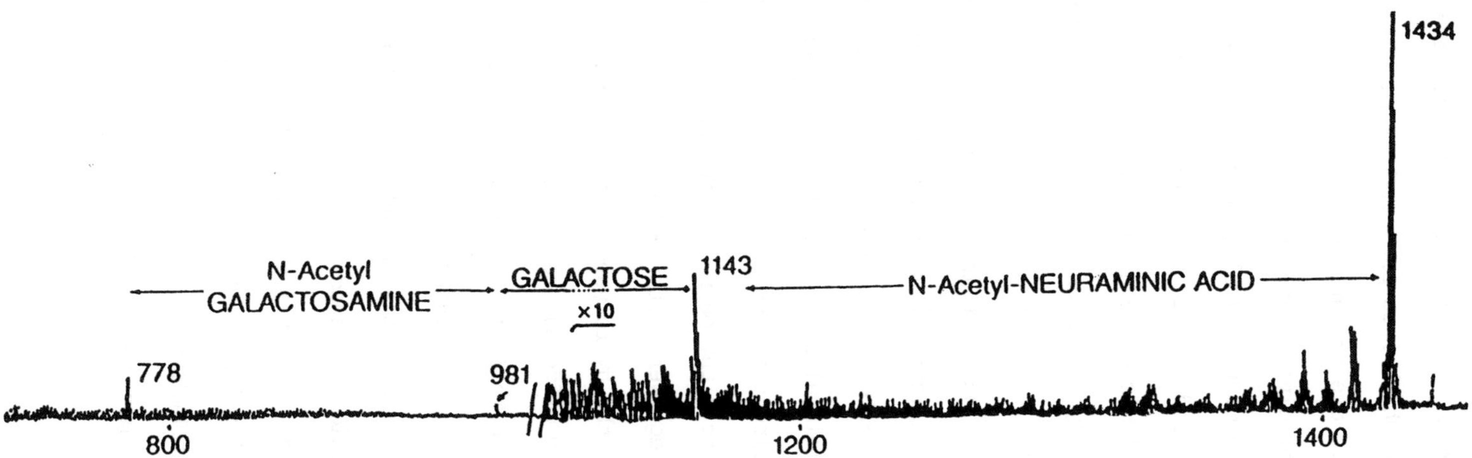

Glycopeptide analysis by FAB-MS. Gas phase Edman sequencing was used to determine the peptide sequence of IL-2 but could not identify the amino acid at position 3 in the sequence. The N-terminal tryptic peptide was isolated and around 1μg used in direct FABMS analysis to produce this spectrum. The quasimolecular ion at m/z 1434 showed that the peptide was post-translationally modified. The fragment ions at m/z 1143, 981 and 778, corresponding to glycosidic bond cleavages between sugar residues, showed that the modification was a glycosylation with N-acetyl-hexosamido-hexosyl-N-acetylneuraminic acid.

illustrated both above (40,41) in the Interleukin 2 research, and by our studies (43) on Erythropoietin which contains a single O-glycan attachment site at Ser126. Tryptic digestion of Erythropoietin, followed by HPLC purification afforded the tryptic glycopeptide spanning residues 117-140. FABMS of this material gave signals at m/z 2412, 2121, 1830, and 1696 corresponding to the expected peptide plus NeuAc2HexHexNAc, NeuAcHexHexNAc, HexHexNAc and HexNAc respectively. It was not, however, clear whether some of the smaller ions were fragment ions or molecular ions. Further experiments, including HPLC separations of the individual glycopeptides and FAB-MS of permethylated samples (prepared by treating the glycopeptide with NaOH/methyl iodide and extracting the permethylated oligosaccharides into chloroform) revealed that the signals at m/z 2121 and 1830 were attributable to both molecular ions and fragment ions of larger components, and that m/z 1696 was a fragment ion.

Often, it may not be necessary to examine glycopeptides, and the most appropriate strategy begins with release of the carbohydrates from the glycoprotein via elimination or reductive elimination. Permethylation of the purified, released oligosaccharides provides samples suitable for analysis by FABMS and GCMS. Alternatively, we recommend an extension of our screening strategy for O-glycans described earlier as a simple procedure for rapidly obtaining samples suitable for sequencing. Peracetylated oligosaccharides obtained at the end of the screening protocol are then permethylated i.e. the acetyl groups are replaced by methyl groups by Hakomori permethylation. This approach circumvents purification at the elimination stage since salts etc are removed from the peracetylated sample by extraction and the products are sufficiently clean for direct permethylation. Permethylated O-glycans afford abundant molecular ions at high sensitivity (100 pmol) and display well defined fragmentation patterns resulting from A-type cleavages (42) at the aminosugar residues. Sequences, branching and sites of fucosylation and sialylation are defined by the fragment ions. Complex mixtures containing up to fifty or more components are amenable to direct FAB analysis (44). Molecular ions can be unambiguously assigned when complex mixtures are examined but fragment ions cannot necessarily be attributed to individual molecular ions. In some instances linked scanning may be helpful in the assignment of fragment ions in a mixture. Usually it is more productive to perform HPLC prior to FABMS if sequence ambiguities remain. Tritium labelling at the initial reduction step provides a useful means of monitoring the separation, while FABMS can then be used to analyse the interesting fractions. Because of the sensitivity of analysis of permethylated samples by FABMS the method lends itself to the screening of HPLC eluents for minor components with retention times very similar to those of more abundant relatives. The effective use of this procedure is illustrated by our research on the O-linked oligosaccharides of human glycophorin (45). After reductive elimination of human glycophorin the released oligosaccharides were permethylated and subjected to HPLC. The major fraction was shown by FABMS to have the correct molecular weight for the expected disialylated tetrasaccharide (m/z 1234). FABMS monitoring of the trailing edge of this peak revealed an unexpected molecular ion at m/z 1595 corresponding to a novel trisialylated pentasaccharide of composition NeuAc3HexHexNAc.

Often it is not necessary to exhaustively purify a mixture of permethylated oligosaccharides in order to sequence the individual components by FABMS. The data in the spectra of simple mixtures can frequently be unambiguous (see, for example, our studies on Leukosialin

(46). In some instances, HPLC purification of the peracetylated sample, coupled with FABMS analysis of the eluted peaks, is a useful first step prior to permethylation. Peracetylated samples often give sequence data additional to that present in spectra of permethylated samples, via less favourable cleavages at hexosyl residues. There are, however, a number of advantages associated with the permethyl derivative : very high sensitivity, easily interpretable spectra with few cleavage pathways and, most importantly, applicability to both FABMS and classical methylation analysis procedures, allowing parallel experiments which furnish sequence and linkage data.

Permethyl derivatives of N-glycans yield FAB spectra containing abundant fragment ions resulting from A-type cleavage at each amino sugar residue. Thus high mannose structures cleave at the chitobiose core to yield fragment ions whose masses define the number of mannosyl residues present, whilst complex structures afford fragment ions from each of the branches, in addition to chitobiose cleavage ions. The branch derived fragment ions define sites of sialylation and fucosylation, presence or absence of bisecting GlcNAc, branching patterns and presence and extent of lactosamine repeat regions (47-51). Sulphate or phosphate functional groups are also revealed by characteristic non-reducing fragment ions. The fragmentation pattern afforded by a permethylated sample is very reproducible. Hence, a comparison of data obtained from two related glycoproteins, e.g. natural and recombinant, will reveal whether the non-reducing structures are similar or, indeed, identical (52). In the high mass region of the spectrum, provided sufficient sensitivity is maintained, chitobiose cleavage ions and molecular ions for the intact oligosaccharides or glycopeptides are observed (N-glycans can be examined

Figure 15:

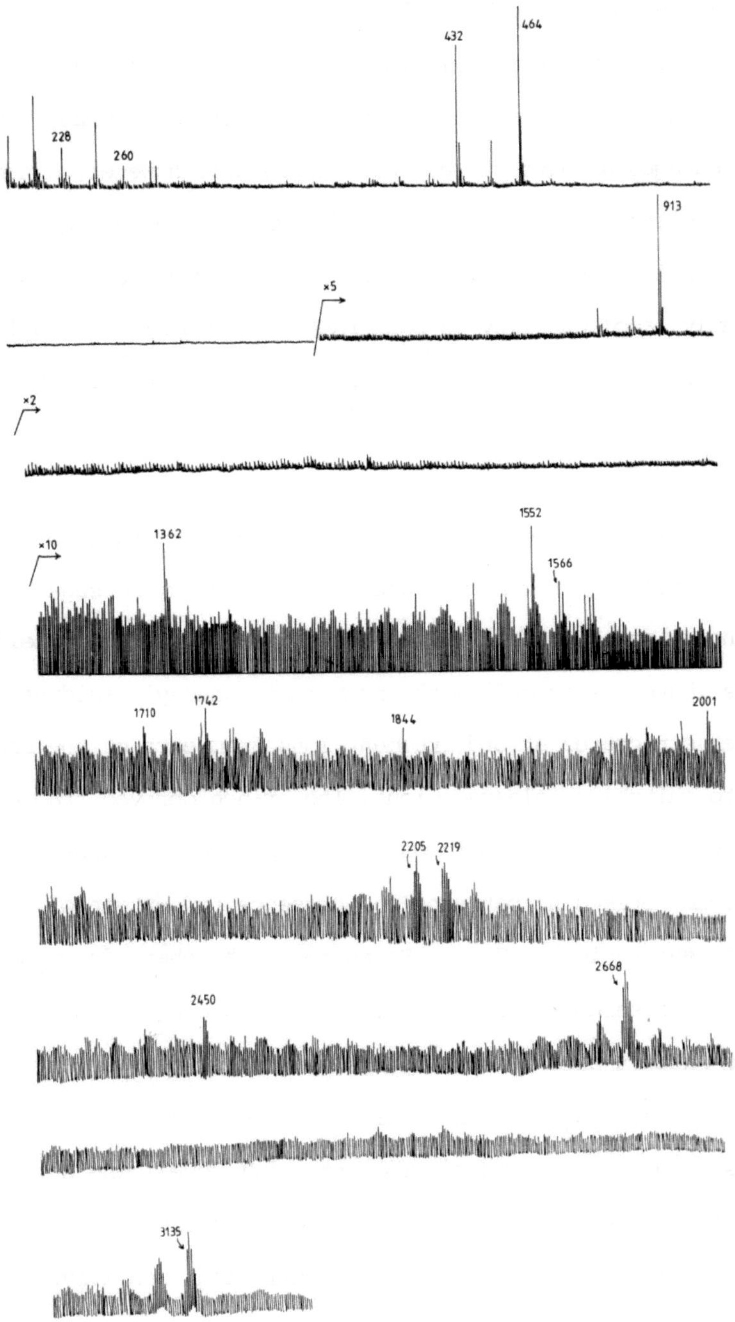

FAB mass spectrum of a permethylated sample of desialylated tri-antennary saccharides from Erythropoietin. The molecular ion is at m/z 3135. Major A-type fragment ions occur at m/z 464, 913 and 2668 corresponding to HexHexNAc, Hex2HexNAc2 and Hex7HexNAc5 respectively.

as their permethylated derivatives either as the free sugars released by hydrazine or N-glycanase or as glycopeptides obtained by proteolytic digestion) . These define the number and types of branches attached to the core, the extent of fucosylation of the reducing end GlcNAc residue and the mass of the aglycone (if present). Mild acid hydrolysis of the sample prior to permethylation can bring larger branched triantennary and tetraantennary structures into view on the mass spectrum by removing bulky N-acetylneuraminic acid residues. For example, in our work on Erythropoietin (52), we were not able to detect the molecular ions of sialylated tri-antennary saccharides containing one lactosaminyl repeat. In contrast, FAB-MS of the saccharides after desialylation afforded an abundant molecular ion at m/z 3135 corresponding to Fuc1Hex7HexNAc6 as the reduced permethylated derivative (Figure 15).

The use of derivatisation to direct fragmentation in FABMS of carbohydrates has been very successful. However, in some molecules, especially highly branched ones, the favoured cleavages may not provide sufficient data for complete sequencing. To overcome this problem we have devised a strategy in which FABMS analyses of chemical hydrolyses are carried out whilst the reaction is proceeding (53). The sample is first permethylated and checked by FABMS to establish whether methylation is reasonably complete. If so, the permethylated sample is subjected to a time course methanolysis during which aliquots are removed at frequent intervals and FAB spectra acquired. The masses of the molecular ions of the hydrolytic fragments define the number of free hydroxyl groups present and these data are used to assign sequence and branching patterns. For example, hydrolytic removal of one or more residues without the generation of free hydroxyl groups indicates that those residues are at the reducing end. Similarly, the production of two free hydroxyl groups indicates the hydrolysis of two different branches. FABMS is an ideal procedure for monitoring the glycosidase reactions which are an integral part of classical carbohydrate sequencing methods. Comparison of spectra acquired before and after digestion with exoglycosidases, endoglycosidases or mild acid will reveal the extent and sites of trimming thereby providing important structural information. For example, FABMS of permethylated N-glycans isolated from the cell surface of acute myelogenous leukemia cells revealed, via a diagnostic signal at m/z 1274, the presence of sialylated branches containing lactosamine repeats. Digestion of the N-glycans with an alpha 2-3 specific neuraminidase yielded a product which, after permethylation, no longer afforded the m/z 1274 signal in the FAB spectrum, demonstrating that the elongated branches contained exclusively alpha 2-3 linked sialic acid (36).

Many of the strategies we have developed for sequencing O- and N-linked glycans are also applicable to biochemical investigations into how such chains are biosynthesised and degraded. The most useful methodology for such studies is the peracetylation/extraction protocol (described above) which allows the rapid analysis of oligosaccharides originally present as part of a very impure sample containing other biopolymers and also high levels of salt. Thus it is possible to monitor the progress of a biological reaction, if one or more of the reactants or products is an oligosaccharide or glycoconjugate, by peracetylation of aliquots at suitable time intervals followed by FABMS of chloroform soluble material (54).

Figure 16:

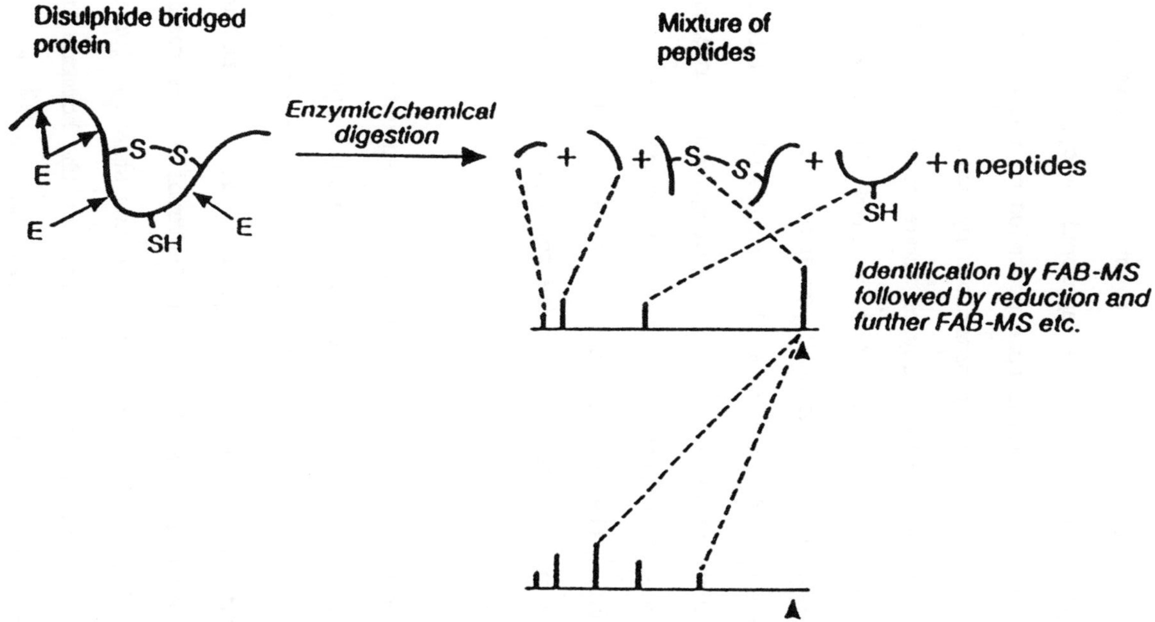

S-S Bridge Assignment using FAB mapping strategy. The polypeptides are cleaved at points initially between the potentially bridged cysteine residues using a suitable method - usually pepsin digestion. The FAB mapping approach is then used to search for any disulphide-bridged peptides. These are characterized by their unique masses, and the interpretation is confirmed by, for example, reduction in thioglycerol or dithiothreitol followed by rerunning the HM-FABMS spectrum. Inter S-S bridged peptides collapse to give two single peaks at lower masses corresponding to the individual peptide components, and a intra S-S bridge (where there is no peptide bond cleavage between cysteins) can simply be diagnosed by a shift of 2 mass units as the disulphide is converted to a dithiole.

S-S Bridge Assignment

One of the most important post translational events for many proteins is the formation of disulphide bridges between appropriate cysteine residues in the primary protein sequence. Whilst some proteins do not contain cysteine and are folded into active conformations by other intramolecular forces, those that contain S-S bridges rely on these covalent bonds for the maintenance of active conformations. Thus, we know that cleavage of S-S bridges without changing the amino acid sequence will destroy enzymic activity, and the stability and kinetics will be altered by single site replacement of an S-S cysteine by site directed mutagenesis. Problems associated with low specific activity in recombinant protein engineering can sometimes be related to the authenticity of the S-S bridges in the product since scrambled or incorrect S-S bridge configurations lead invariably to a functionally defective product.

Based on the FAB mapping procedures described earlier, an entirely new strategy for the important task of S-S bridge assignment in proteins has been devised (Fig 16)(55). This final stage of protein primary structure determination has traditionally been very difficult, and long exhaustive procedures raise the possibility of disulphide "scrambling" or reshuffling by reduction and re-oxidation of S-S bridges. The new strategy for S-S bridge assignment was first demonstrated with work on the hormone Insulin (55), a particularly difficult S-S bridged protein because of the presence of two adjacent cysteine residues in the primary sequence which are difficult to cleave. This problem was solved by repetitive steps of Edman degradation of the peptide digest mixture, monitoring the removal of the Cys residues by examining the FAB spectra of the remaining peptides (55). In disulphide analysis, the relevant signals are normally at the high-mass

Figure 17:

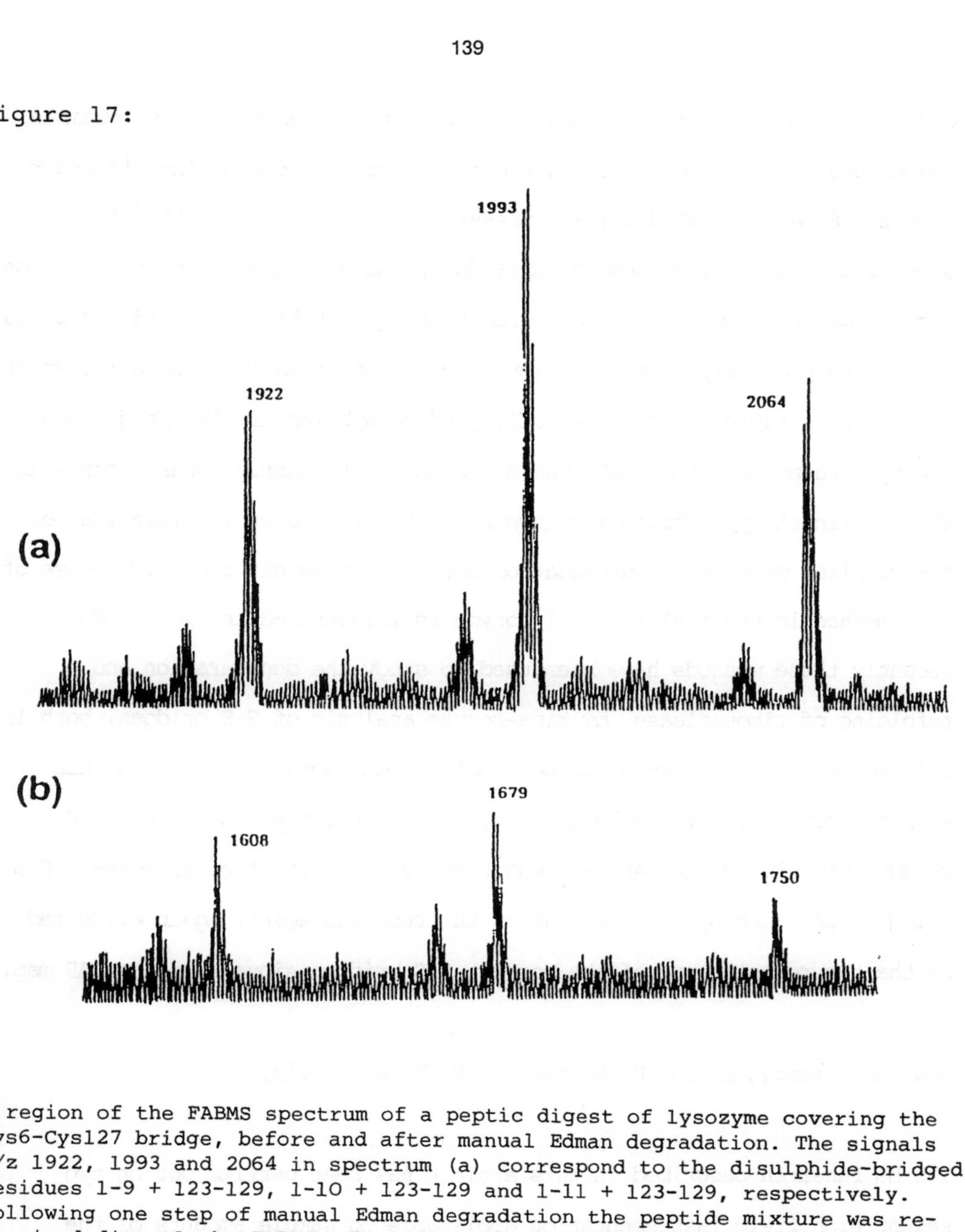

A region of the FABMS spectrum of a peptic digest of lysozyme covering the Cys6-Cys127 bridge, before and after manual Edman degradation. The signals m/z 1922, 1993 and 2064 in spectrum (a) correspond to the disulphide-bridged residues 1-9 + 123-129, 1-10 + 123-129 and 1-11 + 123-129, respectively. Following one step of manual Edman degradation the peptide mixture was re-examined directly by FABMS. The signals in spectrum (b) (m/z 1608, 1679 and 1750) correspond to the cleavage of Trp 123 and Lys 1 (186 and 128 mass units, respectively; 314 in total) from the two N termini of each disulphide-bridged peptide. The S-S bridge itself was not affected by the alkaline pH of Edman degradation. (H.R. Morris and P. Pucci).

end of the spectrum (emphasising the need for High Mass-MS). For example, two of the signals from an unfractionated peptic digest of insulin occur at m/z 2560 and m/z 2788 . Their masses and the observation that the signals at these m/z values collapse following a single S-S bond reduction step allow the signals to be assigned to fragments B1-11 + A1-13 and B1-13 + A1-13 respectively. Since the method does not normally require isolation steps, it is rapid and the possibility of scrambling is minimised. In any event, with proper treatment, the S-S bridges are fairly robust, those in the Insulin study withstanding tryptic, chymotryptic or V8 digestion, or the alkaline pH of Edman degradation for as many as six cycles. The use of this method in the analysis of lysozyme is illustrated in Fig 17. More recently these methods have been used to study the denaturation and refolding of ribonuclease, by time-course analysis of S-S bridges, both in the presence and absence of protein disulphide isomerase (56). In this work the scrambled ribonuclease, diagnosed by the presence of new S-S bridged signals in the FAB map, was then refolded to give the correct S-S bridged structure by slow removal of the reducing agent, again evidenced by the the reappearance of the correct S-S bridged signals in the FAB map.

New Instrumentation for High Mass and High Sensitivity

The MS research described in this article has been carried out on two sector high field instrumentation applicable to sample amounts of the order of 100 picomoles to 1 nanomole and molecular weights up to 6,000 daltons. The original high field concept developed by us has now been extended to larger geometry instruments allowing up to 15,000 dalton basic mass range, and a modern VGZAB 2SE for example can produce molecular weight data on 1-10 picomoles of polypeptide up to 10,000 daltons in mass with an accuracy far greater than achievable on time of flight (TOF) instrumentation. Whilst many have looked towards the development of 4 sector instrumentation to solve the next generation of problems in molecular biology, our own feeling is that a quantum leap in sensitivity is of more importance. A vast array of problems which cannot currently be addressed awaits the development of the femtomole age of high mass mass spectrometry. This is now a realistic goal by the combination of high field magnet and array detector technology, and indeed several devices have been built to date in an attempt to create better sensitivity. The problem is that they allow visualisation of only some 4 or 5% of the mass range at any one time making the efficient screening of genuine unknowns via a survey scan quite unrealistic. This is especially true when one considers the probable lifetime of very small sample quantities during any ionisation procedure - there would simply be insufficient time to cover a reasonable mass range. To solve these problems, in 1987 we suggested a new instrument (now under construction by VG Analytical) which will be directly applicable to the next generation of molecular biology problems. The instrument will have a minimum basic mass range (full accelerating volts) of 12,000 daltons and the array detector, which will incorporate a post accelerator, will allow a minimum 25% and more probably a 50% picture of the mass range, encompassing the complete survey scan within 4 seconds. A second detector will allow 10,000 resolving power for examination of, for example, partial reduction of S-S bridged signals. Most importantly, with these specifications the instrument will allow detection of just 10 femtomoles of biopolymer up to 10,000 daltons in mass, providing a genuine quantum leap in the effectiveness of biopolymer research and analysis. The instrument will aso be available as a 4 sector version (in fact this will now be the first one to be built) to allow daughter ion analysis with increased sensitivity.

Conclusion

High mass fast atom bombardment mass spectrometric techniques have been developed to provide powerful and versatile methods for solving modern protein and glycoprotein structural problems. The procedure is often complementary to classical Edman chemistry, but in addition gives the protein chemist the ability to observe and assign, for example, blocking groups at the N-terminus or heterogeneity at the C-terminus of the molecule. In the analysis of both recombinant and native proteins and glycoproteins, FAB mapping procedures can be used to provide sequence data, detect errors of translation, mutation, insertion, deletion and, importantly, to identify sites of glycosylation or other post - translational modification, to determine the basic carbohydrate structure and to definitively assign S-S bridges. The MS techniques described in this article, based on conventional two-sector double focussing mass spectrometry, are already sensitive ones, applicable to sample sizes of the order of 1 pmol to 1 nmol for molecular weight determination. We have produced realistic specifications for a new generation of high field instruments with post acceleration array detection of ions which should allow detection and measurement of the order of 10 femtomoles of biopolymer, and these ideas have been taken up by a major instrument manufacturer to be put into practice. These methods promise to revolutionise protein and carbohydrate chemistry and biochemistry over the next decade.

Acknowledgements

The authors wish to thank the Medical Research Council U.K. and NATO for financial support.

REFERENCES

1. S. Magnusson, L. Sottrup-Jensen, T.E. Petersen, H.R. Morris, and A. Dell, FEBS Lett., 44:189. (1974).

2. H.R. Morris, A. Dell, T.E. Petersen, L. Sottrup-Jensen and S. Magnusson, Biochem. J., 153:663. (1976).

3. H.R. Morris, M.R. Thompson, D.T. Osuga, A.I. Ahmed, S.M. Chan J.R. Vandenheede and R.E. Feeney, J. Biol. Chem., 253: 5155. (1978).

4. H.R. Morris, G.W. Taylor, P.J. Piper, and J.R. Tippins, Nature, 285: 104 (1980).

5. P. Petrilli, P. Pucci, H.R. Morris, and F. Addeo, Biochem.Biophys.Res.Commun., 140: 28 .(1986).

6. A.J. Geddes, G.N. Graham, H.R. Morris, F. Lucas, M. Barber and W.A. Wolstenholme,Biochem.J., 114: 695.(1969).

7. H.R. Morris, D.H. Williams, G.C. Midwinter, and B.S. Hartley, Biochem.J., 141:701. (1974).

8. W.V. Shaw, L.C. Packman, B.D. Burleigh, A. Dell, H.R. Morris and B.S. Hartley,Nature, 282: 870.(1979).

9. J. Hughes, T.W. Smith, H.W. Kosterlitz, L.A. Fothergill, B.A. Morgan and H.R. Morris,Nature, 258:577. (1975).

10. H.R. Morris, A. Dell, A.E. Banner, S. Evans, R. McDowell and D.

Hazelby,"Extended Performance Using a High Field Magnet Mass Spectrometer",Proceedings of 25th Conf. of Mass Spectrom. and Allied Topics, Washington D.C., pp 73-75 (1977).

11. H.R. Morris, A. Dell, R.A. McDowell, Biomed.Mass Spectrom., 8: 463. (1981).

12. R.D. Macfarlane, and D.F. Torgerson, Science, 191: 920. (1976).

13. M. Barber, R.S. Bordoli, R.D. Sedgwick,and A.N. Tyler, J.Chem.Soc.Chem.Commun., 325. (1981).

14. H.R. Morris, M. Panico, M. Barber, R.S. Bordoli, R.D. Sedgwick and A.N. Tyler, Biochem.Biophys.Res.Commun.,101: 623. (1981).

15. D.H. Williams, C.V. Bradley, S. Santikarn and G. Bojesen, Biochem.J., 201: 105. (1982).

16. H.R. Morris, A. Dell, R.A. McDowell, M. Panico and G.W. Taylor, Mass Spectrometry of Large Molecules (S. Facchetti ,ed.), Elsevier, p.111 and p.127 (1985).

17. H.R. Morris, A. Dell, M. Judkins, R.A. McDowell, M. Panico and G.W. Taylor, Pure Appl. Chem., 54: 267. (1981).

18. K. Biemann, B.W. Gibson, W. F. Mathews and H. Pang, Mass Spectrometry in the Health and Life Sciences,(A.L. Burlingame and N. Castagnoli Jr, eds),Elsevier,,,p.239. (1986).

19. H.R. Morris, A. Dell, M. Panico and R.A. McDowell, Mass Spectrometry in the Health and Life Sciences, (A.L. Burlingame and N. Castagnoli Jr, eds),Elsevier, p.363. (1986).

20. H.R. Morris, M. Panico, and G.W. Taylor, Biochem.Biophys.Res.Commun.,117: 299.(1983).

21. S. Lemaire, L. Chouinard, A. Denis, M. Panico and H.R. Morris, Biochem.Biophys.Res.Commun.,105: 51.(1982).

22. H.R. Morris, M. Panico, and A.T. Etienne "Peptide and Protein

Sequencing by Mass Spectrometry", Proceedings of the 31st Conf. on Mass Spectrometry and Allied Topics, Boston MA, pp683-685 (1983).

23. H.R. Morris, G.W. Taylor, M. Panico, A. Dell, A.T. Etienne, R.A. McDowell and M.B. Judkins, Methods in Protein Sequence Analysis,(M. Elzinga,ed), Humana Press, New Jersey, p.243. (1982).

24. B.W. Gibson, L. Poulter, D.H. Williams and J.E. Maggio, J.Biol.Chem., 261: 5341.(1986).

25. T.J. Knott, S.C. Rall, T.L. Innerarity et al, Science, 230: 37 (1985).

26. T.J. Knott, R.J. Pease, L.M. Powell et al, Nature, 323: 734. (1986).

27. F.M. Greer, H.R. Morris, T. Fallon and S.J. Brewer "FAB Mapping Protein Structural Determination of Recombinant Human Interferon- ", Proceedings of the 35th Conference on Mass Spectrometry and Allied Topics, Denver, Colorado, pp942-943, (1987).

28. H.R. Morris and A. Dell, Biochem.J., 149: 754. (1975).

29. F.M. Greer, H.R. Morris, J. Forstrom and D. Lyons, Biomed.Environ.Mass Spectrom., in press.

30. H.R. Morris, M. Panico, A. Karplus, P.E. Lloyd and B. Riniker, Nature,308:746 (1984).

31. H.R. Morris, M. Panico, A.T. Etienne, J.R. Tippins, S.I. Girgis and I. MacIntyre, Nature, 308:746.(1984).

32. C. Kleanthous, P.M. Cullis and W.V. Shaw, Biochemistry,24:5307(1985).

33. H. Erdjument, D.A. Lane, H. Ireland, M. Panico, V. Di Marzo, I. Blench and H.R. Morris, J.Biol.Chem.,262:1381.(1987).

34. H. Erdjument, D.A. Lane, M. Panico, V. Di Marzo and H.R. Morris, J.Biol.Chem.,263:5589.(1988).

35. D.M. Carlson, J.Biol.Chem.,243:616.(1968).

36. A. Dell, J.E. Thomas-Oates, M.E. Rogers, and P.R. Tiller, Biochimie, in press.

37. A. Dell, J.E. Oates and C.E. Ballou, "Fast Atom Bombardment Mass Spectrometry of High Mass Oligosaccharides and Glycoconjugates", Proceedings of the 7th International Symposium on Glycoconjugates, Lund, Sweden, pp137-138 (1983).

38.S. Naik, J.E. Oates, A. Dell, G.W. Taylor, P.M. Dey and J.B. Pridham, Biochem.Biophys.Res.Commun.,132:1 (1985).

39. S.A. Carr and G.D. Roberts, Analyt.Biochem.,157:396. (1986).

40. R.J. Robb, R.M. Kutny, M. Panico, H.R. Morris, W.F. DeGrado and V. Chowdhry, Biochem.Biophys.Res.Commun, 116:1049.(1983).

41. R.J. Robb, R. M. Kutny, M. Panico, H.R. Morris and V. Chowdry, Proc.Natl.Acad.Sci.U.S.A., 81: 6486.(1984)

42. A. Dell, Adv. in Carbohydr. Chem. and Biochem,45:19. (1987).

43. Sasaki H., Bothner B., A. Dell and M. Fukuda, submitted for publication.

44. A. Dell, M.E. Rogers and Fukuda M., unpublished work.

45. M. Fukuda, M. Lauffenberger, H. Sasaki, M.E. Rogers and A. Dell, J.Biol.Chem.,262: 11952. (1987).

46. M. Fukuda, S.R. Carlsson, J.C. Klock and A. Dell, J.Biol.Chem.,261:12796. (1986).

47. M. Fukuda, A. Dell, and M.N. Fukuda, J.Biol.Chem.,259:4782 (1984).

48. M.Fukuda, B. Bothner, B. Ramasamooj, A. Dell, P.R. Tiller, A. Varki and J.C. Klock, J.Biol.Chem., 260:12957 (1985).

49. M. Fukuda, A. Dell, J.E. Oates and M.N. Fukuda, J.Biol.Chem.,259:8260. (1984).

50. M.N. Fukuda, A. Dell, J.E. Oates and M. Fukuda, J.Biol.Chem.,260: 6623. (1985)

51. J.E. Oates, A. Dell, M. Fukuda and M.N. Fukuda, Carbohydr.Res.,141:149 (1985).

52. H. Sasaki, B. Bothner, A. Dell and M. Fukuda, J.Biol.Chem.,262: 12059. (1987).

53. A. Dell, N.H. Carman, P.R. Tiller and J.E. Thomas-Oates, Biomed.Environ.Mass Spectrom., in press.

54. C. Barker, A. Dell, M. Rogers, J.A. Alhadeff and B. Winchester, Biochem.J., in press.

55. H.R. Morris and P. Pucci, Biochem.Biophys.Res.Commun.126: 1122. (1985).

56. H.R. Morris, P. Pucci, M. Panico and G. Marino, J.Biol.Chem., submitted for publication.

IDENTIFICATION OF ACTIVE-SITE IN A NEUROTOXIC SNAKE VENOM BY AFFINITY LABELLING AND STATE-OF-THE-ART TANDEM MASS SPECTROMETRY TECHNOLOGY

Thaiya Krishnamurthy & Marguerite E. Brooks*
Research Directorate
US Army Chemical Research, Development and Engineering Center
Aberdeen Proving Ground, MD 21010-5423 USA

Donald F. Hunt & Jeffrey Shabanowitz
University of Virginia
Charlottesville, VA 21903 USA

Shuian Chen & Terry Lee
Beckman Research Institute of City of Hope
Duarte, CA 91010 USA

ABSTRACT

α-Bungarotoxin, a post-synaptic neurotoxic snake venom, is a protein of molecular weight 7,994 Daltons. The sequences of all the amino acids in the molecule were deduced using nanomole quantities of the toxin and utilizing the state-of-the-art fast atom bombardment tandem mass spectrometric techniques. Active-site of the venom was identified by affinity labelling followed by the sequencing of the labelled molecule.

INTRODUCTION

Snake venoms are complex mixtures of substances with various types of toxic properties. Some are neurotoxic and others are cardiotoxic, cytotoxic, or ion channel blockers. Active components of several snake venoms have been deduced to be proteins.[1] A large number of venoms have been studied by conventional sequencing methods.[1] However, only a limited account of their structures, specific active-sites, and immunogenic regions, required for the development of specific methods of detection and antibody production, etc., is available.[1]

Based on a few chemical modification studies and comparison of the sequences of amino acids of neurotoxic and cytotoxic snake venoms, the following conclusions have been made.[1,2] Some amino acids have been identified with the specific activity of the molecule and hence are termed as functional invariants. Whereas, some other amino acid moieties in each venom do not seem to play any major roles with the toxicity of the molecule. However, they are required for the general folding, conformation, and disulfide pairing of the toxins essential for their specific binding. Hence, these are known as structural invariants.[1,2] Most of the above are assumed without any experimental data.

Recently, we devised a strategy based on the tandem mass spectrometric techniques [3,4] in combination with the HPLC and enzymatic methods in order to experimentally determine the sequences of the amino acids and to detect active-sites in several snake venoms. We have applied it to verify the amino acid sequences in α-bungarotoxin reported in the literature.[5,6] The intact toxin

U. P. Schlunegger (Ed.)
Biologically Active Molecules

was subjected to reduction and carboxymethylation followed by specific enzymatic cleavages. The products were separated over a microbore HPLC column and lyophilized. The fractions and their corresponding methyl esters were ionized under fast atom bombardment conditions and their corresponding masses were determined. Daughter mass spectra of selected fractions containing oligopeptides and their methyl esters provided their corresponding sequences of amino acids present in the molecule. Logical compilation of all the observed sequences provided the total sequence data of the toxin. Active-sites of α-bungarotoxin were then detected by labelling the purified venom with azo compounds, followed by the identification of the labelled-site (active-site) from the sequences of the labelled molecule. Identification of the oligopeptide with the label, resulting from the enzymatic cleavage of the labelled α-bungarotoxin, could be applied for the specific detection of the venom in samples, development of specific antibody, etc.

This type of approach to determine the sequences of amino acids in a peptide or protein is helpful even when blocked and unusual amino acids are present, when conventional sequencing methods fail. In addition, the amino acid sequences of the entire molecule even with affinity labels can be determined rapidly and more importantly with accuracy. The determination of the amino acid sequences in proteinaceous toxins and venoms, accurately and rapidly, is the initial and vital step to be undertaken for the detection of such larger biomolecules in samples and treatment of affected personnel.

EXPERIMENTAL METHODS

The chemicals mentioned in this paper, including α-bungarotoxin, were purchased either from Sigma Chemicals (St. Louis, MO) or Pierce Chemicals (Rockford, IL). All solvents were of HPLC grade and bought from Burdick and Jackson (Muskegon, MI). Trypsin and chymotrypsin were purchased from Sigma Chemicals (St. Louis, MO) and Staphylococcus Aureus Protease V8 from Miles Laboratories (Elkart, IN).

The HPLC analysis was performed on an Applied Biosystems Model 130A Separation System (Foster City, CA) using the microbore RP-300 Aquapore Octyl column (2.1 mm x 3 cm, 10 μM particles) from Brownlee Labs, Rainin Instruments Co. (Woburn, MA). Mass spectra and daughter spectra were recorded either on a triple quadrupole mass spectrometer [3] or on a tandem quadrupole Fourier Transform (QFT) mass spectrometer.[4].

REDUCTION AND CARBOXYMETHYLATION OF α-BUNGAROTOXIN: The venom (50 nMoles) in a polypropylene vial (2 ml) was dissolved in 50 nMolar ammonium bicarbonate (500 μl) and treated with 100 nMole/ul dithiothreiotol solution (50 μl) and left at 37°C for 30 minutes with occasional agitation. The reaction mixture was then treated with iodoacetic acid solution in freshly degassed 50 nMole ammonium bicarbonate (50 nMole/μl; 50 μl). . The pH of the mixture was then adjusted to 8.6 using solid ammonium bicarbonate. The vial was sealed in a nitrogen atmosphere and heated in dark at 37°C for 30 minutes. The mixture was lyophilized overnight and the residue was treated with water (200 μl) and lyophilized in order to remove the excess buffer. The residue was dissolved in 5% acetic acid (300 μl) and divided into three equal portions in 3 separate polypropylene vials (2 ml) and lyophilized again. The residues containing ca. 17 nMoles were used for digesting with different enzymes.

DERIVATIZATION 4-VINYLPYRIDINE: 1-5 nMole of α-bungarotoxin was dissolved in 50 ul of buffer containing the following in specified concentrations: tris-HCl (0.25M), EDTA (1.0 mM), guanidine.HCl (6.0 M). The solution was then treated with 2.5 ul of acqueous mercaptoethanol(1.43M) and incubated under argon for two hrs in ambient temperature. The product was mixed with 2 ul of neat 4-vinylpyridine and incubated for an additional two hours under the same conditions. The solution was stored under at -27°C until separation over a C-4 HPLC microbore column.

PREPARATION AND PURIFICATION OF LABELLED α-BUNGAROTOXIN: A solution containing α-bungarotoxin (6 nMole) in borate buffer (16 μl) was treated with 4-azidophenylglyoxal (15 nMole) or 4-fluoro-3-nitrophenylazide (15 nMole) and incubated at 37°C in the dark for 2 hours. The reaction products were then cooled and loaded on a microbore Vydac C-18 or C-4 column in order to purify the labelled molecule. The column was then eluted with 0-90% acetonitrile (0.1% TFA)/0.1% aqueous TFA in for minutes with a flow rate of 0.25ml/min. Column effluent was monitored at 214nM.

PROTEOLYTIC OR ENZYMATIC DIGESTION: All three portions of residues obtained as above were dissolved in 50nM ammonium bicarbonate solution (100 μl) treated with 5 ug of enzyme (1-2% w/w enzyme/protein ratio) and incubated at 37°C for 24 hrs. The three enzymes used in this investigation were trypsin, chymotrypsin and S.aureus. The digests were then lyophilized in order to arrest the reaction and the residues were dissolved in 35-45 μl of 5% aqueous acetic acid and separated by HPLC. Similarly the carboxymethylated labelled α-bungarotoxin was also subjected to cleavage with trypsin.

HPLC SEPARATION OF OLIGOPEPTIDES FROM THE ENZYMATIC DIGESTS: The product in 5% acetic acid (35-50 ul) was loaded on an Aquapore column. A 40 minute linear gradient of 0% to 60% acetonitrile containing 0.085% trifluoroacetic acid in 0.1% aqueous trifluoroacetic acid was employed to separate the mixtures. Fractions containing each peptide were collected and divided into three equal portions and all portions were lyophilized. Fractions containing the mixtures of oligopeptides were further chromatographed over a C-18 column using the above elution conditions.

PREPARATION OF METHYL ESTERS: A solution of 2N methanolic HCl was prepared by adding 16ml of acetyl chloride dropwise to 100 ml of methanol. After standing at ambient temperature for 5 minutes, 200 ul of the methanolic HCl solution was added to each of the HPLC fractions in order to convert them into their corresponding methyl esters. The reaction was kept at room temperature for 2 hrs with occasional shaking and excess reagent was removed by lyophilization.

MASS SPECTROMETRY: Mass spectra were recorded on a triple quadrupole mass spectrometer assembled with components from Finnigan-MAT (San Jose, CA) [3] by using 1 μl of a 5% acetic acid solution containing the peptide at 0.5-1.0 nMole level to 1 μl of thioglycerol on a gold-plated, stainless steel 2 mm probe tip. The peptides were sputtered out of the liquid matrix by bombardment with 6-8 kev Cs^+ ions generated with a cesium ion gun [3], mounted directly on the source. The masses of the peptides in each sample were measured by operating Q3 in the rf and dc mode and the other quadrupoles in the rf mode only in order to focus all the ions up to m/z 1,800 Daltons into Q3. The masses of the ions were determined by scanning Q3 between the mass ranges, 400-1,800 Daltons.

When the mass of the oligopeptides in the enzymatic digests exceeded 1,800 Daltons, the products were analyzed in the tandem quadrupole Fourier Transform mass spectrometer.[4] Samples were ionized using a liquid matrix and a cesium gun (Antek, Palo Alto, CA).[3] Experimental details are described elsewhere in detail.[4]

TANDEM MASS SPECTROMETRY: An aliquot (1 μl) of oligopeptide solutions were used to determine their corresponding sequences. The sample was ionized as before. The molecular ion of interest was chosen by Q1 and sent to collision chamber(Q2), which was filled with argon to a pressure of 5 mTorr. The fragment ions formed as a result of collision in Q2, which was operated in the rf mode only, were focused into Q3. The Q3 were scanned between mass ranges m/z50-1,800 at a scan rate of 200 Daltons/second. The collisionally activated dissociation (CAD) mass spectra were obtained by summing up the acquired six scans. Thus, the CAD (daughter) spectra of the fragments present in all of the HPLC fractions were obtained.

Some of the MS/MS spectra of larger oligopeptides (mw >1,800 Daltons) were recorded in a VG Zab-SE/Zab-SE 4 sector double focussing tandem mass spectrometer with the courtesy of VG Instruments, Manchester, UK.

RESULTS AND DISCUSSION

The amino acid sequences obtained by the conventional sequencing method (Edman Degradation) indicated the molecular weight of the intact protein to be 7,994 Daltons.[5,6] The amino acid composition [5] of α-bungarotoxin is listed in Table 1. There are no potential glycosylation sites in the reported sequence. Ten cysteine moieties and 5 disulfide bonds were also reported.[5,6] The theoretical isoelectric point is 7.97 and the estimated half life is approximately 30 minutes.[1]

The entire approach pursued by us for the determination of sequences of amino acids in α-bungarotoxin is illustrated in the scheme. The disulfide bridges present at the cysteine moieties of the intact α-bungarotoxin (50 nMoles) were reduced by treating the protein with dithiothreitol, followed by the introduction of carboxymethyl groups at each of the SH groups by treating the mixture with iodoacetic acid. The molecular weight of the carboxymethylated toxin was calculated to be 8,584 Daltons.

The products were divided into three equal parts each containing 17 nMoles) and each of them was subjected to enzymatic cleavage. One portion was digested with trypsin in order to cleave the bonds specifically at lysine and arginine. The bonds at glutamic and aspartic acids were cleaved by digesting the second portion with S.aureus. Similarly, the bonds at leucine, tyrosine, tryptophan, phenylalanine were cleaved by digesting the third portion with chymotrypsin. The digests were incubated overnight at 37°C and lyophilized to stop the cleavage and remove the enzyme from the products. The residues were dissolved (separately) with 40-45 μl of 5% aqueous acetic acid, loaded on a microbore RP-300 column and the components were separated, as specified in the experimental section. The LC chromatogram of the trypsin digest is shown in Figure 1. Eluent corresponding to each peak was manually collected in polypropylene tubes, and lyophilized. Each residue was treated with 5% acetic acid (30 μl), agitated with Vortex and centrifuged. Two 10 μl portions from each vial were collected in separate vials and lyophilized. One of the two portions were used for the

SCHEME

INTACT TOXIN(SNAKE VENOM)

|

Reduction and Carboxymethylation

|

Enzymatic Cleavage

|

HPLC Separation of Products

|

Collection and Lyophilization of Fractions.

|

Conversion of Aliquots of Fractions to Methyl Esters

|

Fast Atom Bombardment Mass Spectra of All Fractions and the Corresponding Methyl Esters.

|

Collisionally Activated Dissociation (Daughter or MS/MS) Mass Spectra of Selected Fractions and Derivatives

|

Determination of Sequences in Individual Peptides.

|

Compilation of Sequences of All Fragments.

|

Sequences of Amino Acids in

α-BUNGAROTOXIN

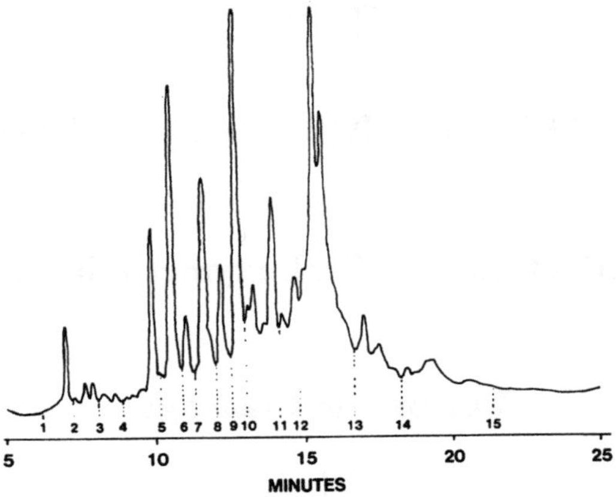

Figure 1. HPLC SEPARATION OF TRYPTIC DIGEST OF CARBOXYMETHYLATED α-BUNGAROTOXIN

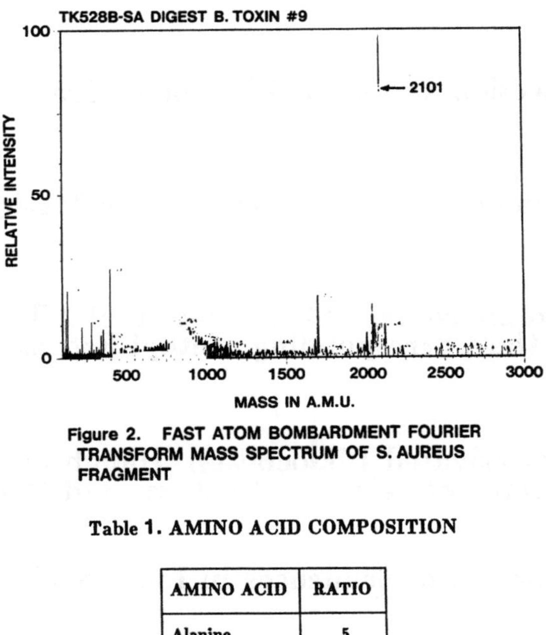

Figure 2. FAST ATOM BOMBARDMENT FOURIER TRANSFORM MASS SPECTRUM OF S. AUREUS FRAGMENT

Table 1. AMINO ACID COMPOSITION

AMINO ACID	RATIO
Alanine	5
Arginine	3
Asparagine	2
Aspartic acid	2
Cysteine	10
Glutamine	1
Glutamic acid	4
Glycine	4
Histidine	2
Isoleucine	2
Leucine	2
Lysine	6
Methionine	1
Phenylalanine	1
Proline	8
Serine	6
Threonine	7
Tryptophan	1
Tyrosine	2
Valine	5

preparation of corresponding methyl esters of polypeptides and the other for recording the higher mass daughter spectral data, etc. Methyl esters of all the fractions were prepared by treating the residues with methanolic HCl. Main beam spectra of all of the fractions and their corresponding methyl esters were recorded. The molecular weights of most of the fractions were found to exceed the mass range of the triple quadrupole mass spectrometer. The masses of the molecular ions of those fragments were determined from their Fourier Transform mass spectra (Figure 2).

The fractions containing the oligopeptides exceeding the mass range m/z 1,800 Daltons were subjected to a second enzymatic cleavage in order to cleave them into smaller fragments, suitable to analyze by triple quadrupole mass spectrometer. Enzymatic digestions were carried as stated above and the products were lyophilized. Main beam mass spectra of all of the products were recorded. Similarly, their corresponding methyl esters were also digested with adequate enzymes, lyophilized, and treated with methanolic HCl in order to esterify the new oligopeptides formed during the second enzymatic digestion. The masses of the methyl esters were also determined from their main beam mass spectra.

The CAD (daughter) mass spectra of all of the oligopeptides with mass ranges less than 1,800 Daltons, and their corresponding methyl esters, were recorded using the triple quadrupole mass spectrometer. Daughter spectrum of a methyl ester of a tryptic fragment (m/z 1450) is shown in Figure 3. The CAD spectra of some of the larger fragements prior to a second enzymatic cleavage were recorded on a 4 sector magnetic high resolution tandem mass spectrometer (Zab SE/Zab SE) with the courtesy of VG Instruments, Manchester, UK. The CAD spectrum of a S.aureus fragment thus obtained is illustrated in Figure 4.

The sequences of the oligopeptides were determined from the B-type and Y"-type [7] ions observed in their corresponding daughter ion spectra. The sequences of the peptides thus deduced are listed in Table 2. Logical compilation of all of the observed sequences led to the total sequences of the amino acids in α-bungarotoxin (Figure 5). It should be noted that distinction between amino acids with the same molecular weights and fragmentation patterns under CAD conditions is not possible by the tandem mass spectrometric technique. Thus, the presence of leucine and isoleucine in the bungarotoxin could not be differentiated from our findings, so it was made from the reported sequences [5,6], which are also illustrated in Figure 5. The sequences of the amino acids, 9-11 (from the n-terminus), deduced by us differed from one of the reported sequences.[5,6] The sequences of 12 amino acids from the n-terminus of the intact toxin were also determined using Associated Biosystems (ABI) gas-phase amino acid sequencer. Our results on the sequences of 9-11 moieties corraborated with the observations made by Ohta et al.[6] However, the sequences of amino acids observed [6] from 67-72 amino acid moieties differed from the latter (Figure 5)

4-Azidophenylglyoxal(APG)

m.w. 175

4-Fluoro-3-nitrophenylazide(FNPA)

VVELGCAATCPSK

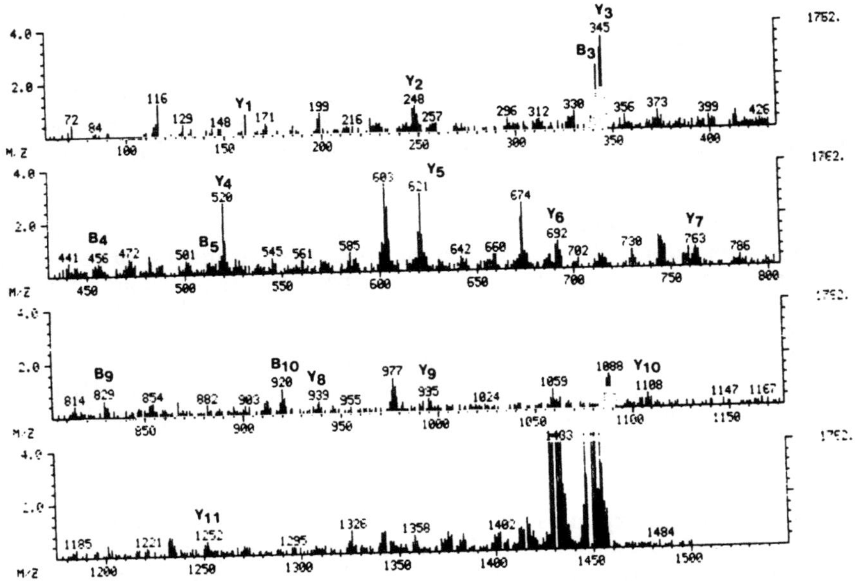

Figure 3. COLLISIONALLY ACTIVATED DISSOCIATION MASS (DAUGHTER) SPECTRUM OF METHYL ESTER OF TRYPTIC DIGEST FRAGMENT (m/z 1450).

113	212	373	510	611	712	783	884	971	1068	1181	1268	1339	1438	1539	1700	1797	1894	1951	2099
Ile--	Val--	Cys--	His--	Thr--	Thr--	Ala--	Thr--	Ser--	Pro	Ile--	Ser--	Ala--	Val--	Thr--	Cys--	Pro--	Pro--	Gly--	Glu
2099	1986	1887	1726	1589	1488	1387	1316	1215	1128	1031	918	831	760	661	560	399	302	205	148

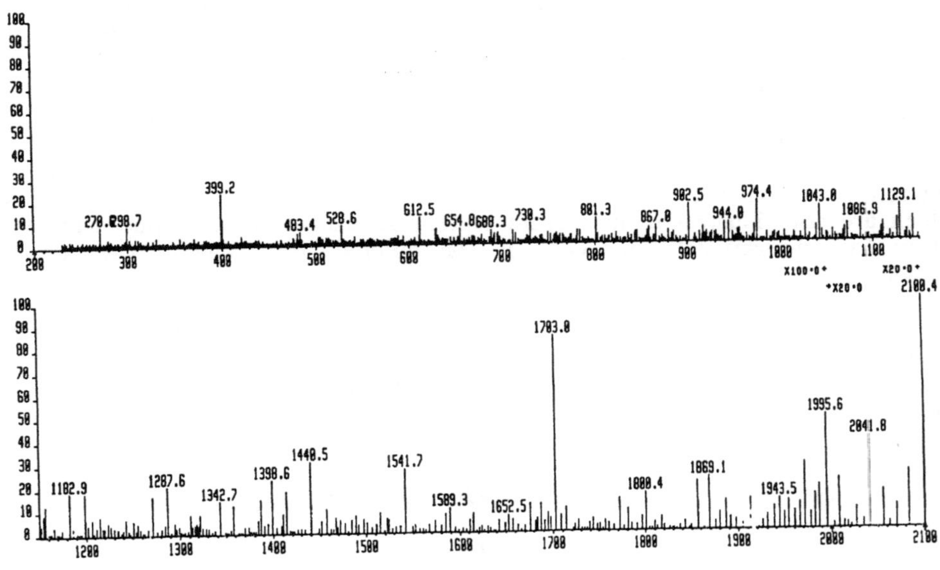

Figure 4. DAUGHTER SPECTRUM OF S. AUREUS DIGEST (m/z 2100)

Table 2. PROTEOLYTIC DIGESTION PRODUCTS

ENZYME	m/z* CALCULATED	m/z* OBSERVED	SEQUENCES OF FRAGMENTS
Chymotrypsin	2328	2328	IVCHTTATSPISAVTCPPGENL
	2653	2653	IVCHTTATSPISAVTCPPGENLCY
	513	513	CDAF
	1136	1136	CSSRGKVVEL
	2308	2308	MWCDAFCSSRGKVVELGCAA
	2461	2461	CSSRGKVVELGCAATCPKKPY
	2404	2404	EEVTCCSTDKCNHPPKRQPG
Trypsin	2809	2809	IVCHTTATSPISAVTCPPGENLCYR
	1322	1322	MWCDAFCSSR
	1393	1393	VVELGCAATCPSK
	1578	1578	VVELGCAATCPSKGK
	1619	1619	KPYEEVTCCSTDK
	2922	2922	DAFCSSRGKVVELGCAATCPSKKPYE
S.aureus	2101	2101	IVCHTTATSPISAVTCPPGE
	2671	2671	NLCYRKMWCDAFCSSRGKVVE
	1712	1712	LGCAATCPSKKPYE
	2668	2668	NLCTRKMTCDFCSSRGKVVE

```
          10          20          30
          |           |           |
IVCHTTATSPISAVTCPPGENLCYRKMWCD
          ---
          40          50          60
          |           |           |
AFCSSRGKVVELGCAATCPSKKPYEEVTCC

          70
          |
STDKCNHPPKRQPG
```

Figure 5. SEQUENCES OF AMINO ACIDS IN alpha-BUNGAROTOXIN

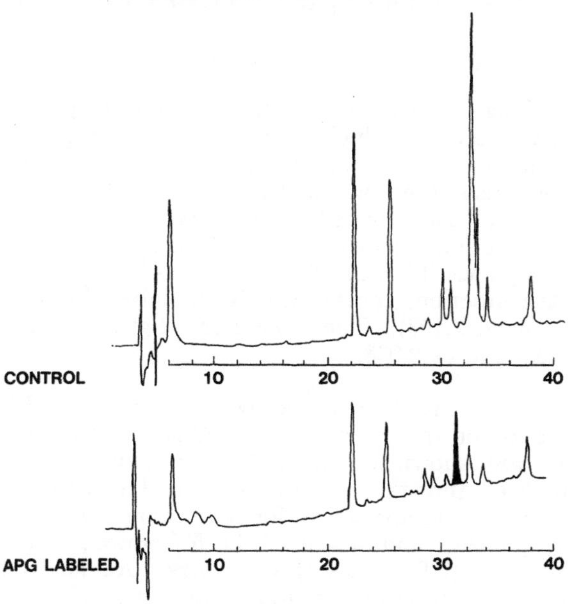

Figure 6. SEPARATION OF TRYPTIC DIGEST OF APG LABELLED α-BUNGAROTOXIN

m.w. 182

One nanomole of α-bungarotoxin in borate buffer was digested with three nanomoles of p-azidophenylglyoxal(APG), for modifying arginine specifically. [8] The product would result from the addition of glyoxal to bungarotoxin at the arginine moiety followed by the formation of schiff's base. The product was reduced, derivatized with vinylpyridine and purified by chromatographic separation over a microbore C-18 HPLC column and lyophilized. The residue in ammonium bicarbonate (50nMole) buffer was cleaved with trypsin (1 μg). The digest containing the mixture of oligopeptides was separated over a C-4 microbore column (Figure 6). Similarly, one nanomole of the unlabelled α-bungarotoxin as the control was also digested with trypsin and separated (Figure 6). One of the fractions eluting at 31.5 minutes was found only in the preteolytic product of the labeled toxin and not in the control. The collected sample was lyophilized and dissolved in 5% aqueous acetic acid (μl) and analyzed by tandem quadrupole Fourier Transforn Mass Spectrometry. The QFT spectrum indicated only one ion at m/z3209. All other fractions were also analyzed similarly in order to detect any other lableled fraction present, if any. None was detected in others.

Computer analysis using PEPED [9] software was performed in order to identify the residue present in the α-bungarotoxin with m/z 3052 (3209-175+18). The analysis indicated such a residue was not present in the toxin. However, when the decomposition of the azido group into the nitrene followed by the conversion into an amino group (3209-131; 175- 18-28+2) was also incorporated and the resulting mass m/z 3078 analyzed the sequences of the amino acids in the labeled fraction was deduced to be:

IVCHTTATSPISAVTCPPGENLCYRK

The AGP label was incorporated into the Arg-25.

Similarly, labelling of the toxin with 4-fluoro-3-nitrophenylazide was carried out. One of the LC fractions, which is highlighted in Figure 7A, was found to contain an oligopeptide with molecular weight of 2,193 (retention time, 14.5 minutes). There were also other ions in this fraction, all of which were observed in the tryptic digest of the unlabelled molecule. This fraction was further purified by chromatographing over a C-18 column (0-40% 0.1% TFA in acetonitrile/0.085% aqueous TFA in 40 minutes). All fractions were analyzed as usual and one of the fractions contained only the labelled oligopeptide (Retention time, 18.9 minutes; m.w. 2,193). The molecular weight of this fragment without the label was calculated to be 2,038.

Computer analysis using PEPED[9] software was performed in order to acquire the projected sequence of the residue (m.w. 2,038) in the bungarotoxin. The projected sequences are shown in Table 3. The solution containing the labelled fragment was lyophilized and the residue was treated with methanolic HCl (50 μl) and left at ambient temperature for two hours with occasional agitation. The product was lyophilized and dissolved in 5% aqueous acetic acid (5 μl) and analyzed using QFT mass spectrometry. An ion with m/z 2,248 was observed indicating the addition of four ester methyl groups to the labelled fragment. The last two entries in Table 3. with masses 2,041 were ruled out based on the fact both would have added on 6 methyl groups on methylation. The second and third fragments would only have resulted from the S.areous and

TABLE 3. PROJECTED SEQUENCES OF AMINO ACIDS IN LABELLED FRAGMENT

RESIDUE	MASS	AMINO ACID SEQUENCE	# of FREE COOH GROUPS
34-52	2038.3	SSRGKVVELGCAATCPSKK	4
38-55	2040.4	KVVELGCAATCPSKKPYE	4
6-24	2041.3	TATSPISAVTCPPGENLCY	4
39-56	2041.3	VVELGCAATCPSKKPYEE	6
40-57	2041.3	VELGCAATCPSKKPYEEV	6

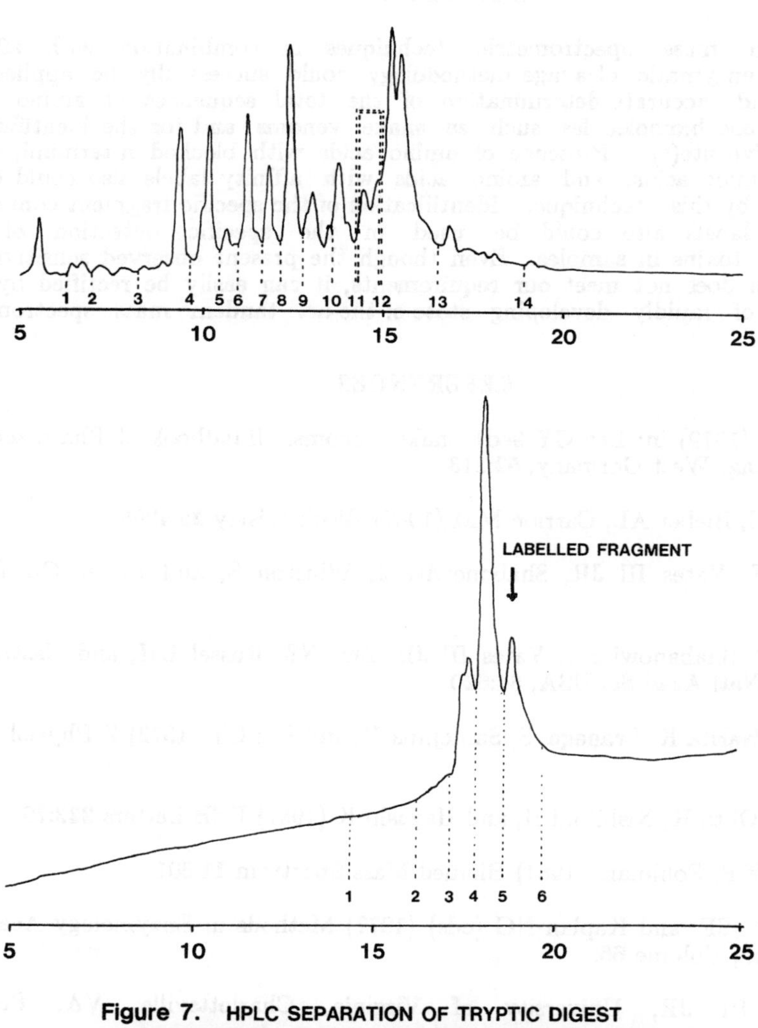

Figure 7. HPLC SEPARATION OF TRYPTIC DIGEST OF CARBOXYMETHYLATED LABELLED α-BUNGAROTOXIN

chymotrypsin cleavages of α-bungarotoxin, respectively. In addition, all except the first entry have a higher molecular weight than the observed one. Hence, it was concluded that the first projected sequence of the labelled fragment shown in Table 3. would be a more adequate one. The FNPA label was added to the Lys-51.

Two (of the) active-sites involved in the binding of α-bungarotoxin with the acetylcholine receptor have been identified to be Arg-25 and Lys-51. This observation is not in disagreement with the reported results.[2] Investigation leading to the identification of all of the binding sites present in the α-bungarotoxin is underway. This approach could also be applied for the detection of α-bungarotoxin in samples.

CONCLUSION

Tandem mass spectrometric techniques in combination with affinity labelling and enzymatic cleavage methodology could successfully be applied for the rapid and accurate determination of the total sequences of amino acids in larger toxic biomolecules such as snake venoms and for the identification of their active-site(s). Presence of amino acids with blocked n-terminii, other uncommon amino acids, and amino acids with affinity labels also could easily be detected by this technique. Identification of the specific fragment containing the affinity labels also could be used in the specific detection of the proteinaceous toxins in samples. Even though the present observed sensitivity in this approach does not meet our requirements, it can easily be rectified by the application of rapidly developing state-of-the-art tandem mass spectrometric technology.

REFERENCES.

1. Low BW (1979) in: Lee CY 9ed) Snake Venoms. Handbook of Pharmacology, Springer Verlag, West Germany, **52**:213

2. Borron JC, Bieber AL, Carrion MM (1987) Biochemistry **26**:4295

3. Hunt DF, Yates III JR, Shabonowitz J, Winston S, and Hauer CR (1986) **83**:6233

4. Hunt DF, Shabanowitz J, Yates III JR, Zhu NZ, Russel DH, and Castro ME (1987) Proc Natl Acad Sci USA, **84**:620

5. Mebs D, Narita K, Iwanaga, S, Samejima Y, and Lee CY (1972) Z Physiol Chem **353**: 243-262

6. Ohta M, Ohta K, Nishitani H, and Hayashi K (1987) FEB Letters **222**:79

7. Roepstorff P, Fohlman, (1984) Biomed Mass Spectrom **11**:601

8. Colowick SF and Kaplan NG (eds) (1977) Methods in Enzymology Academic Press, London, Volume 66.

9. Yates III JR, University of Virginia, Charlottsville, VA, Personal Communications, 1988.

PEPTIDE DEFENCE SYSTEMS

by

Dudley H. Williams
University Chemical Laboratory
Lensfield Road,
Cambridge, CB2 1EW

Introduction

Medical practice has benefitted throughout its history from the use of substances produced by living systems other than Homo sapiens. Even natural products which appear to have evolved as toxins find use in medicine. For example, the curare alkaloids, produced by various species of Strychnos and Chondodendron and used as arrow poisons by Indians in the Amazon and Orinoco valleys, are used in small quantities as skeletal muscle relaxants. The search for novel physiologically active substances is an important part of the endeavour to improve clinical practice. In this account, I will describe a number of structures which have been determined in my own research group during the last 10 years. These structures are all peptides, and therefore are of particular interest at a time when pharmaceutical companies are looking at the therapeutic potential of peptides more actively than at any previous time. Additionally, the solution conformations of peptides are now of considerable interest, because a knowledge of these conformations may increase our understanding of molecular recognition phenomena, which provide the functional basis of all living organisms - and hence cause clinically observable phenomena. The account is concerned with two distinct topics - first, the determination of the structures themselves; and second (in two cases), studies of the modes of action of the peptides.

About 25 years ago, it was considered by some that the advent of X-ray crystallography had largely removed the need for the structure determination of natural products by other methods. The main

U. P. Schlunegger (Ed.)
Biologically Active Molecules

traditional method had been chemical degradations and, towards the middle of the 1960's, important help was also obtained from mass spectrometry and nuclear magnetic resonance (nmr). Today, it can be seen that X-ray crystallography has not in fact dominated the scene in the structure determination of natural products. One reason for this is that many materials of natural origin are available only in extremely small quantities. A second reason is that these substances, even when available in larger quantities, are frequently not crystalline. Thirdly, even if crystals are obtained, a structure cannot always be deduced by X-ray methods. Last, and not least, the techniques of mass spectrometry and nuclear magnetic resonance have increased in power over the last 25 years to such an extent that even quite complicated structures can now be worked out relatively rapidly using only small amounts of material. Therefore, I will first illustrate how the structures of medium-sized polar, and physiologically active, molecules are currently deduced. In the second part of the article, I will illustrate how proton nuclear magnetic resonance can be used to establish the molecular basis of action of one such molecule. In the last part of the presentation, I will indicate how a combination of techniques has been used to establish the structures of the first peptide antibiotics to be isolated from a vertebrate, and how a plausible mode of action of these antibiotics can be inferred.

Structure Elucidation

(i) Mass Spectrometric Determination of a Linear Peptide Sequence

Mass spectrometry is a technique which showed much promise for the elucidation of structures in the middle 1960s. It could be shown that compounds of known structure fragmented in a manner that was understandable in terms of energetic concepts that were familiar to organic chemists (1). However, the reverse exercise, that of elucidating the structure of an unknown presented only with a mass spectrum, is much more difficult. Consequently, it has gradually been realised over the years that very often a mass spectrum alone is of limited use in the determination of new and complex structures. Yet mass spectrometry does have several unique attributes as a physical method. First, it allows the determination of molecular weights on extremely small amounts of material (often only one microgram or less is needed). Second, if the substance to be studied is a linear molecule made up of known building blocks, then a great deal of structural information can be obtained from the fragmentation pattern found in the mass spectrum. The latter potential is uniquely realised in the mass spectra of peptides, and this is particularly true since the advent of fast atom bombardment mass spectrometry (FAB MS)(2). In this technique, a few micrograms of an unknown are dissolved in a matrix such as glycerol, and this solution is then bombarded with an atom beam of several keV energy. The result is that highly polar molecules are desorbed into the gas phase, and peptides up to molecular weights of several thousand daltons can be directly examined. In the mass spectra which are obtained, the ions corresponding to the intact unknown are extremely abundant. These ions appear either by virtue of protonation of the sample, to give MH^+ in a positive ion spectrum; or alternatively, in a negative ion spectrum $M\text{-}H^-$ appears by proton loss from the originally neutral sample. Although the fragment ions obtained in the mass spectrum of a peptide are considerably less abundant than the molecular ions, they are sufficiently abundant in many cases to give a partial sequence of the peptide, and less frequently even a full sequence. The manner in which a typical set of fragment ions is formed from a protonated molecular ion is indicated in Scheme 1.

How the technique works in practice can be illustrated by the stucture determination of a peptide, lophyrotomin, isolated from an extract of the caterpillar which grows into the adult sawfly <u>Lophyrotoma interrupta</u> (3). The caterpillar is an agricultural pest in north-eastern Australia because in times of drought it falls from the leaves of eucalypts where it would normally feed. The larvae which accumulate on the ground are then often eaten by cattle and sheep, which as a result may die from liver poisoning within a few days. The toxic principle (named lophyrotomin) in the larvae was shown by amino acid analysis to correspond to a peptide containing the amino acids Ala(1), Phe(1), Val(1), Ile(1), Asp(2), and Glu(2). The proton NMR spectrum indicated the possible presence of a benzoyl group. The positive ion FAB mass spectrum of the toxin is reproduced in Figure 1. The presence of the abundant MH^+ ion at m/z 1040 corresponds to the presence of four carboxy and one primary amide group in the toxin. Four kinds of sequence ions are observed in the spectrum. Those sequence ions which contain the N-terminal portion of the peptide correspond to acylium ions (1), or protonated amide ions (2); (2) carries a charge by virtue of being protonated at one site, and the $CONH_2$ group is generated from a backbone secondary amide by hydrogen rearrangement from the portion lost as a neutral moiety. Ions corresponding to (1) occur at m/z 105, 176, 323, 422, 535, 650, 765, and 894 (and MH^+ at 1040); this indicates the sequence C_6H_5COAlaPheValIleAspAspGluGln, or the related sequence with the C-terminal Gln replaced by the isomeric amino acid carrying the amide on the backbone carbonyl and the carboxy function in the side chain. Ions corresponding to (2) at m/z 667, 782, and 911 (in conjunction with MH^+) confirm the nature of the three amino acids constituting the C-terminal portion.

Those sequence ions which contain the C-terminal portion of the peptide correspond to protonated amine fragments (3), or protonated alkyl ion fragments (4). Each of these fragments is formed via a hydrogen rearrangement from the incipient neutral moiety. Ions corresponding to (3) at m/z 865, 718, 619 and 506 (Figure 1) indicate the sequence XPheValIleY, where the masses of X and Y are in accord with the proposed sequence. Very low abundance ions, corresponding to (4), at m/z 849, 702, 603, and 490 confirm this sequence. Despite the low abundance of this last series of ions, the presence of the attendant isotope peaks, and their relationship to the series based upon (3) (16 mass units lower), leave us in no doubt as to their structural significance.

The spectrum reproduced in Figure 1 was obtained only after traces of sodium and potassium salts had been carefully removed from the toxin by ion exchange chromatography. Prior to the chromatography, the spectrum was more complex and showed peaks in the molecular ion region at m/z 1078
$(M + K^+)$, 1084 $[(M + 2Na^+) - H^+]$, 1100$[(M + Na^+ + K^+) - H^+]$, 1,116$[(M + 2K^+) - H^+]$, 1,122$[(M + 2Na^+ + K^+) - 2H^+]$,
1,138$[(M + 2K^+ + Na^+) - 2H^+]$, and 1,154$[(M + 3K^+) - 2H^+]$.

SCHEME 1

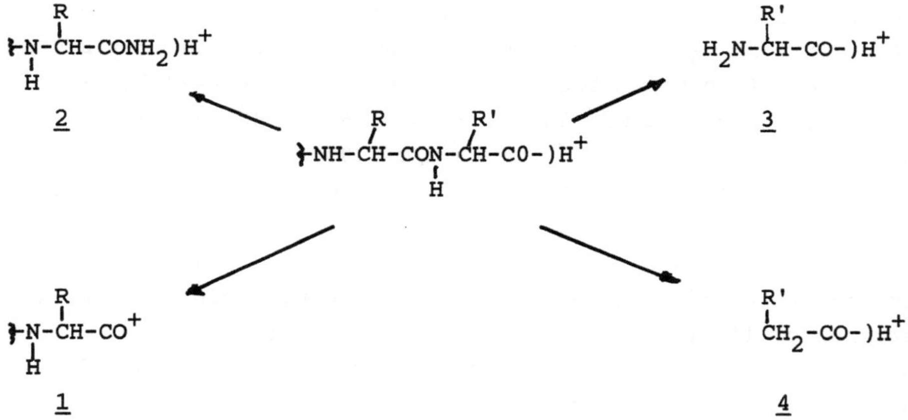

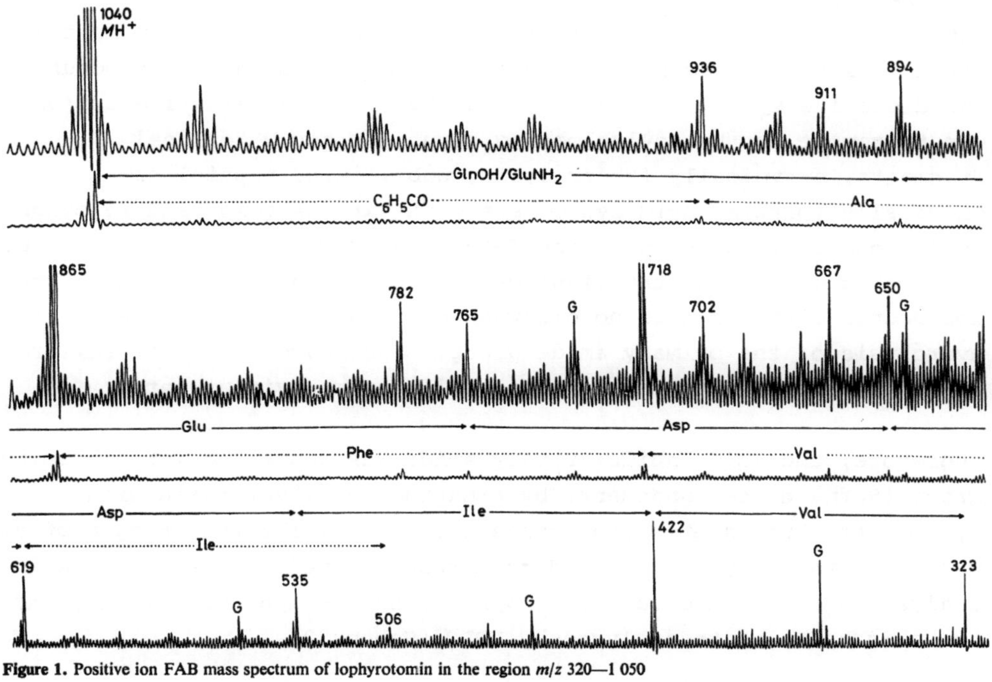

Figure 1. Positive ion FAB mass spectrum of lophyrotomin in the region m/z 320—1 050

A number of the structural conclusions derived from the positive ion FAB spectrum could be confirmed from the negative ion FAB spectrum (Figure 2). This shows an extremely abundant $(M - H)^-$ ion at m/z 1,038. A series of sequence ions, most generally expressed in terms of the anion (5), occur at m/z 338, 437, 550, 665, 780 and 909. These together with the $(M - H)^-$ ion at m/z 1,038 indicate the sequence XValIleAspAspGluGln (or, as before, the C-terminal residue may be iso-Gln).

It remains for us to distinguish between the two possible isomers at the C-terminus of liver toxin; the primary amide group of the C-terminal amino acid could possible be at the α-position (iso-Gln) or at the γ-position (Gln). The amino acids glutamine and isoglutamine can be readily distinguished chromatographically. However, all attempts to release the C-terminal residue enzymically (carboxypeptidase A; carboxypeptidase Y; combinations of several non-specific proteases and aminopeptidases) failed. This is perhaps not surprising since any peptide that is toxic by ingestion must be either reasonably resistant to enzymic hydrolysis which can occur in the digestive system, or quickly absorbed. The alternative approach was to remove the benzoyl blocking group, or any N-terminal fragments, enzymically (thermolysin, subtilisin, papain) or chemically (partial acid hydrolysis with various acids and reaction conditions). In principle, the C-terminal sequence information could then be determined by the classical Edman degradation method. Such approaches either caused no cleavage or resulted in concurrent hydrolysis of the primary amide group. The reasons for the failures of the attempted enzymic cleavages will become evident later.

Therefore, the acid labile primary amide was converted into an amino group (Hofmann rearrangement) by treating the liver toxin with I,I-bis(trifluoroacetoxy)iodobenzene (TIB)(4). The resulting peptide was (i) totally hydrolysed and the products subjected to amino acid analysis and (ii) treated with naphthalene-1-sulphonyl chloride and the acid hydrolysate analysed by HPLC (Scheme 2).

The results from the amino acid analysis (Table 1) clearly show that treatment of acetylated glutamine and liver toxin resulted in (i) the

$R'CONH^-$

<u>5</u>

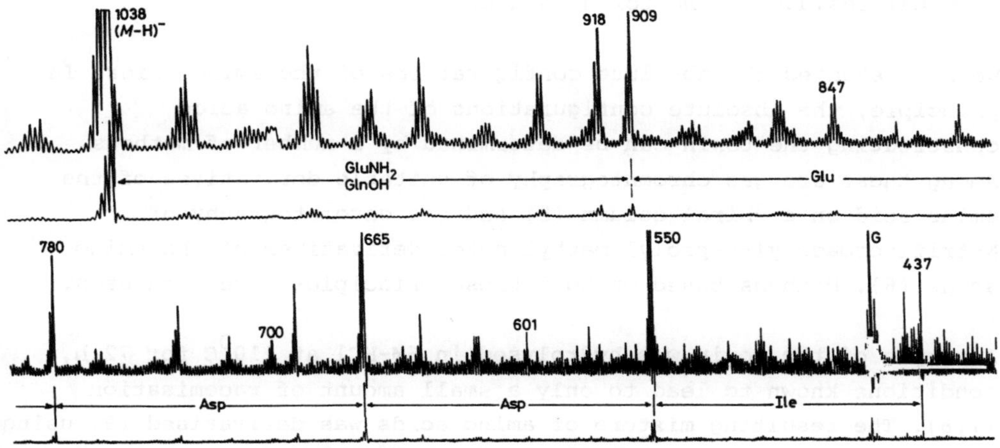

Figure 2. Negative ion FAB mass spectrum of lophyrotomin in the region *m/z* 410—1 050

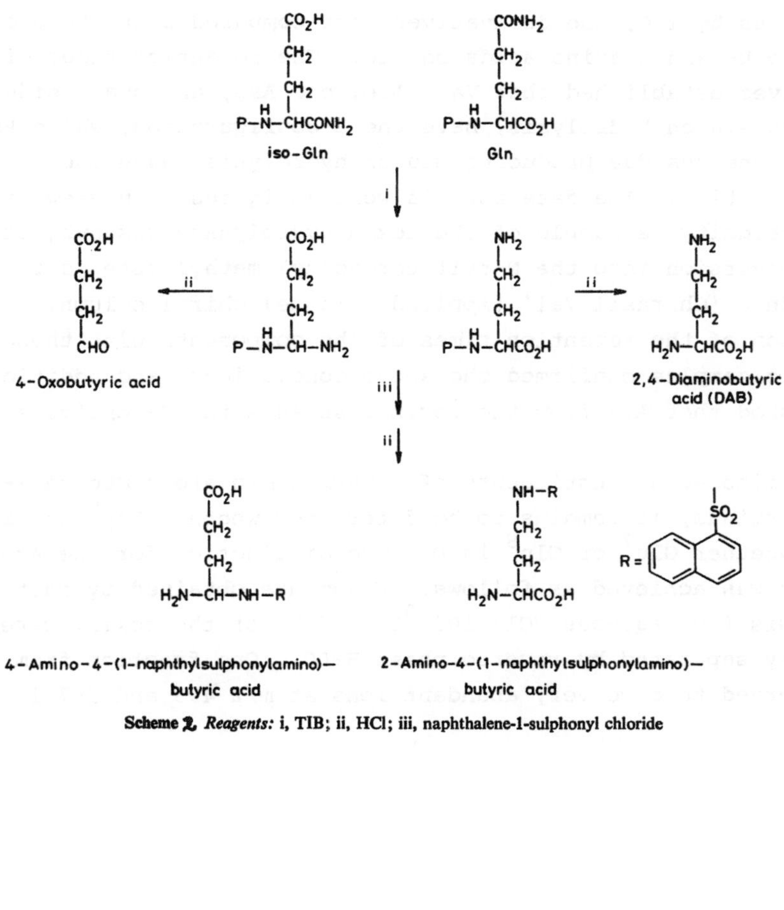

Scheme 2. *Reagents:* i, TIB; ii, HCl; iii, naphthalene-1-sulphonyl chloride

appearance of an amino acid with identical retention time with 2,4-diaminobutyric acid, and (ii) reduction in the amount of glutamic acid obtained after acid hydrolysis. These results indicate that the terminal residue of the toxin is glutamine.

We next studied the absolute configurations of the amino acids. In principle, the absolute configurations of the amino acids constituting the toxin can be determined by a number of methods. Among these are gas chromatography of suitable derivatives of the amino acid on a chiral column (5) and gas chromatography of N-trifluoroacetyl-L-prolyl methyl ester derivatives of the amino acids (6). Methods based on both these principles have been used.

A sample of the toxin was hydrolysed in 6M-HCl at 110^0C for 72 h, conditions known to lead to only a small amount of racemisation (7,8). The resulting mixture of amino acids was derivatised (9) using N-trifluoromethyl-L-propyl chloride prepared by a literature procedure. After partial separation of the resulting mixture of derivatives by TLC, the derivatives were compared with those of authentic D- and L-amino acids on GLC. The retention times of the derivatives established that Val, Ile, one Asp, and one residue producing Glu on hydrolysis, have the L-configuration, while Phe, one Asp, and one residue producing Glu on hydrolysis, have the D-configuration. The data for Ala were ambiguous. In view of the above ambiguity, a sample of the toxin hydrolysate was subjected, after conversion into the N-trifluoroacetyl methyl ester derivatives, to GLC on a 'Chirasil-Val' (Applied Science) chiral column. Comparison of the retention times of the components with those of authentic samples confirmed the above conclusions, and additionally established that Ala from the toxin possesses the D-configuration.

If the amino acid constituents of lophyrotomin are numbered 1-8 from the N-terminus, it remains to be determined whether Asp^5 or Asp^6 is D, and whether Glu^7 or Gln^8 is D. The distinction for the Asp residues was achieved as follows. Fragments obtained by partial acid hydrolysis (10% aqueous HCl; 100 ^{0}C; 1.5 h) of the toxin, were partially separated by reverse phase HPLC. One fraction from HPLC was observed to give very abundant ions at m/z 493 and 247 in the

Table 1. Amino Acid Composition after Treatment with TIB

Amino acid ratio[a]	Starting material				Control
	Ac-Gln	Ac-iso-Gln	2,4-DAB[c]	Liver toxin	Liver Toxin
Ala				1.00	1.00
Phe				0.96	0.91
Val				0.94	1.16
Ile				1.01	0.82
Asp				1.93	1.89
Glu	0.3[b]	0.3[b]		1.19	1.93
2,4-DAB	0.6[b]		1.0[b]	0.89	

[a]As 4-oxo-butyric acid lacks a primary amino group, it does not form a coloured derivative with ninhydrin and thus is not detected in the amino acid analyser. [b]Relative to the amount of the starting material. [c]2,4-DAB, Lys, and Orn are not resolved on this analyser

Table 2. Isotopic Abundance and Relative Amounts of Some D- and L-amino acid derivatives in LT-C and LT-D

	% Isotopic Abundance Relative to $(M - H)^-$ Peak			
	O	+1	+2	D:L Ratio
(i) LT-C				
D-Glu	100	15	4	100:40
L-Glu	100	14		
D-DAB	100	11	4	10:100
L-DAB	100	13	2	
D-Asp	100	19	13	
L-Asp	100	20		80:100
(ii) LT-D				
D-Glu	100	20	3	100:20
L-Glu	100	43	14	
D-DAB	32	100	14	20:100
L-DAB	100	66	8	
D-Asp	100	17	13	100:100
L-Asp	100	17	13	

positive ion FAB mass spectrum. These ions correspond to MH^+ signals from the peptides PheValIleAsp and IleAsp. Support for the assignment of the peak at m/z 493 came from the fragmentation pattern, which established loss of C-terminal Asp [loss of 116 to give ion type (2) and loss of 133 to give ion type (1)]. Additionally, esterification of this peptide with MeOH-HCl gave, by FAB analysis, a diester (MH^+ at m/z 521), and acetylation with Ac_2O-H_2O gave a monoacetyl derivative (MH^+ at m/z 535), as required. The m/z 247 ion also showed the expected increments in these experiments. Following the above analytical experiments, the remainder of the above fraction from HPLC was completely hydrolysed with 6M-HCl, and the resulting mixture of amino acids converted into the corresponding N-trifluoroacetyl isopropyl esters. These derivatives were then subjected to GLC on a chiral 'Chirasil-Val' column. Co-chromatography with authentic derivatives of D- and L-Asp, and D- and L-Phe established that the hydrolysate contains both D- and L-Asp, and D- and L-Phe. The observed D:L Asp ratio was ca 6:1 and the D:L Phe ratio > 10:1. These results confirm that D-Phe occurs in the toxin, and establish that Asp^5 is D. The smaller ratio of D:L Asp (relative to D:L Phe) may be due to the expected limited amount of racemisation of Asp during hydrolysis, and contamination of the fraction with traces of peptides containing both Asp residues.

The absolute configurations of residues 7 and 8 were determined from a complete acid hydrolysate of the crude Hofmann degradation product, previously established is containing 2,4-diaminobutyric acid (DAB) at the C-terminus. As before, N-trifluoroacetyl isopropyl esters of the resulting amino acids were prepared, and these analysed by GC-MS on a 'Chirasil-Val' column. The amino acid derivatives were identified by their retention times as well as their positive negative ion CI-MS, while their absolute configurations were determined by co-chromatography with authentic derivatives. The hydrolysate contained D- and L-Glu in the ratio ca 93:7, and D- and L-DAB in the ratio ca 5:95. Thus the penultimate residue, Glu^7, has the D-configuration.

The presence of D-amino acids in liver toxin strongly indicates that the peptide was generated by a non-ribosomal biosynthetic pathway. Therefore it cannot be assumed that all the aminodicarboxylic acids, D- and L-Asp and D-Glu, form peptide bonds through the α-carboxy groups. The possibility that Asp and/or Glu may be present in the peptide as iso-Asp and/or iso-Glu was investigated by a selective deuteriation procedure, using a published method (10) with the modification that D_2O was used instead of T_2O. The selective deuteration of an amino acid residue containing a free α-carboxy group is based upon oxazolone formation in peptides by the action of acetic anhydride, followed by base-catalysed hydrolytic ring opening (in D_2O). This results in racemization and deuteriation of the α-carbon in the amino acid (Scheme 3).

Liver toxin was treated with acetic anhydride-pyridine in the presence of D_2O (LT-D), and a control sample was not thus treated. Both samples were then subjected to Hofmann degradation and total hydrolysis. The chiralities of the amino acids produced were analysed by GC-MS using negative ion chemical ionization (NH_3), as described earlier; the results are given in Table 2. It can be seen that deuterium was not incorporated into aspartic acid and therefore the toxin does not contain any iso-Asp residues. About 10% of L-DAB racemized to the D-isomer in LT-C, probably during the acid hydrolysis step. From this it can be calculated (using the data in Table 2) that roughly 20-30% of L-DAB in LT-D racemized with deuterium incorporation. This results in an increase in the amount of D-DAB relative to the L-isomer, and in both isomers having unusually abundant +1 isotope peaks. This result confirms that the toxin contains a C-terminal glutamine residue. The data for the glutamic acid are complicated by the fact that in this experiment ca. 30% of the C-terminal Gln did not undergo the Hofmann degradation. The hydrolysis product of LT-C therefore contains ca. 40% of L-Glu relative to D-Glu. The racemization of L-Glu in LT-D produced equal amounts of deuterium labelled D- and L-Gln. The proportion of each that did not undergo Hofmann degradation was hydrolysed to deuterium labelled D- and L-Glu, equal amounts of which supplemented the unlabelled material. Since there was originally ca. five times more unlabelled D-Glu than unlabelled L-Glu, this addition of deuteriated

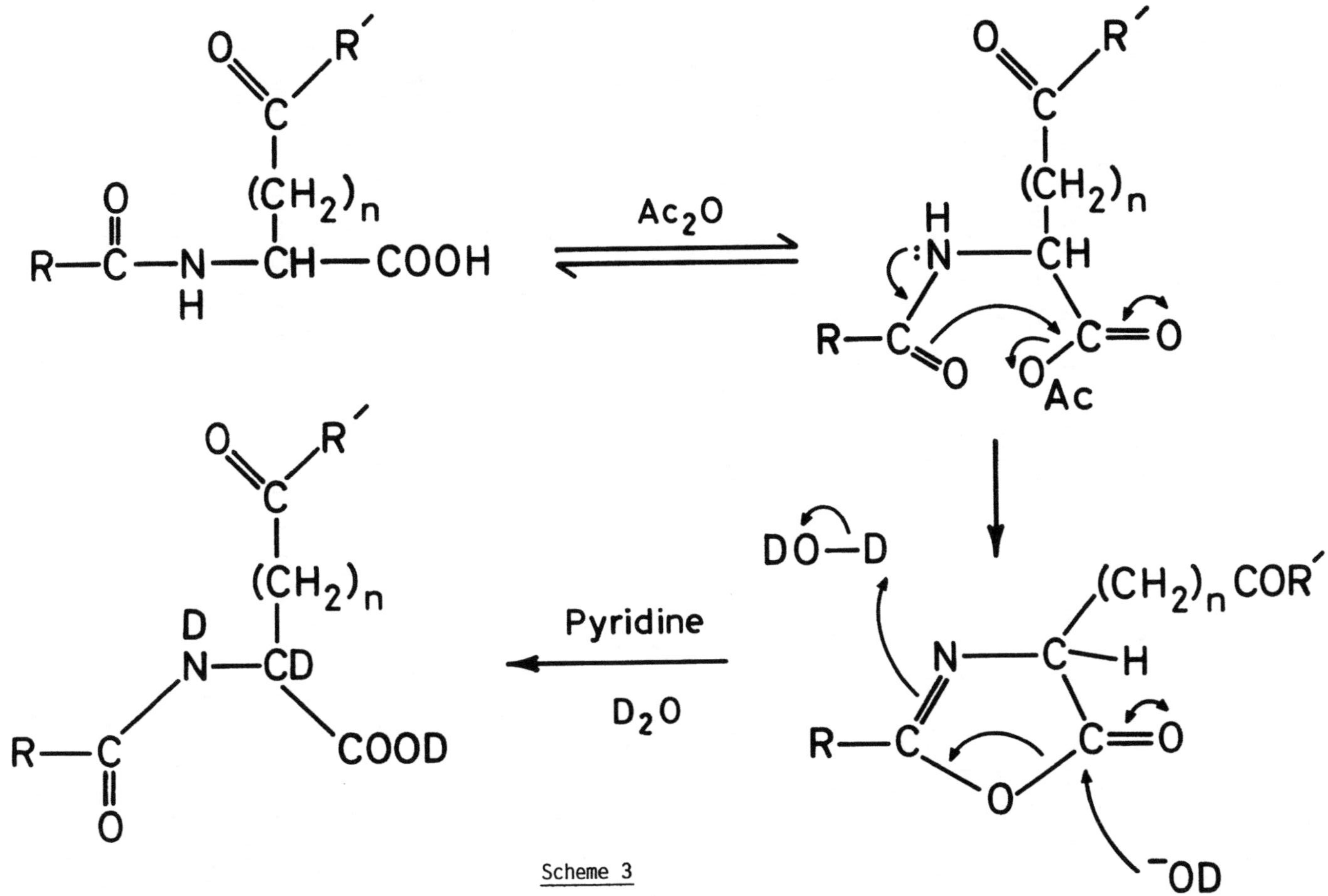

Scheme 3

Glu had a much greater effect on the abundance of the +1 isotope peak of the L-Glu derivative than on that of the D-Glu derivative. It is concluded from these experiments that the toxin does not contain an iso-Glu residue.

The structure of lophyrotomin, a toxic octapeptide from the larvae of an Australian species of sawfly (Lophyrotoma interrupta) is thus C_6H_5CO-(D)-Ala-(D)-Phe-(L)-Val-(L)-Ile-(D)-Asp-(L)-Asp-(D)-Glu-(L)-Gln.

Although the site of action of the toxin in the cell is as yet undetermined, it is noteworthy that the N-terminal region of the peptide is highly hydrophobic, and the C-terminal portion relatively hydrophilic, with several carboxy groups. Thus the N-terminal portion should be readily lipid (membrane) soluble, while the C-terminal portion is suitable to bind metal ions. Thus, it is interesting to speculate that the toxicity may be related to the binding of metal ions within the cell. That the toxin may have evolved as a deterrent to predators of the larvae is indicated by the resistance which will exist to its hydrolysis by the predator. This resistance is conferred by the presence of several D-amino acids and a blocked N-terminus.

The presence of four D-amino acids in a peptide isolated from an animal is, so far as we are aware, without precedent. However, at this stage we cannot preclude the possibility that the peptide, or some or all of its D-amino acid components, is biosynthesised by one or more of the simpler organisms with which the larvae interact.

(ii) Structure Elucidation of Cyclic Peptides

(a) A Monocyclic Peptide Microcystin

Peptides produced by microorganisms are frequently much more complex than those produced by animals. For example, they may be polycyclic, and in these circumstances it is necessary to find some selective method for opening one or more rings in the peptide before a mass spectrometric sequencing strategy can be reliably applied. This was the case in a structural study on a peptide toxin, microcystin, produced by the blue-green algae Microcystis aeruginosa(11). The structure (6) for this peptide was deduced in part by first of all reducing the unsaturated dehydro-alanine unit with sodium borohydride, and then opening the peptide macrocycle by a short hydrolysis with 6NHCl at 110^{0}C for 5 minutes. The linear peptide so obtained could then be sequenced by electron impact mass spectrometry using principles which are rather similar to the ones previously enunciated. However, note that in this case a highly unusual β-amino acid with a long hydrocarbon sidechain is present. Although mass spectrometry can indicate the mass of this residue, it is a totally inappropriate technique to give the full structure of the sidechain. This detailed structure was determined by using high-field proton nmr. First by using decoupling experiments (which can determine the through-bond connectivity of protons); and second by using nuclear Overhauser effects (nOes), which can indicate the spatial proximity of protons. The latter technique is of such enormous power for the determination of structure in solution that the general principle involved is laid out in Scheme 4.

Although the molecular basis for the toxicity for microcystin is not known, it seems reasonable to hypothesise that it has evolved in order to protect the algae from being used by other organisms as a food source. So high is the toxicity of microcystin that there are reports of liver damage in human populations that have drunk water from dams where the algae has bloomed on the dam surface.

(b) A Tricyclic Peptide Moroidin

Peptide toxins in plants are relatively unusual, but we have recently been able to determine the structure of one such toxin, moroidin (7), which appears to be only one component of the toxic principles produced by the Australian bush Laportea moroides(12).

Moroidin showed a λ_{max} at 215 nm and 275 nm in the UV spectrum. It gave a positive test with chlorine-starch-iodine suggesting that it might be a peptide. Amino acid analysis showed Moroidin to contain (molar ratio in parentheses): Glu(1), Gly(1), Val(1), Leu(1), His(1) and Arg(1). Moroidin gave no reaction with Pauly's reagent (sulphanilamide/HCl; $NaNO_2/H_2O$). This test is positive only if an

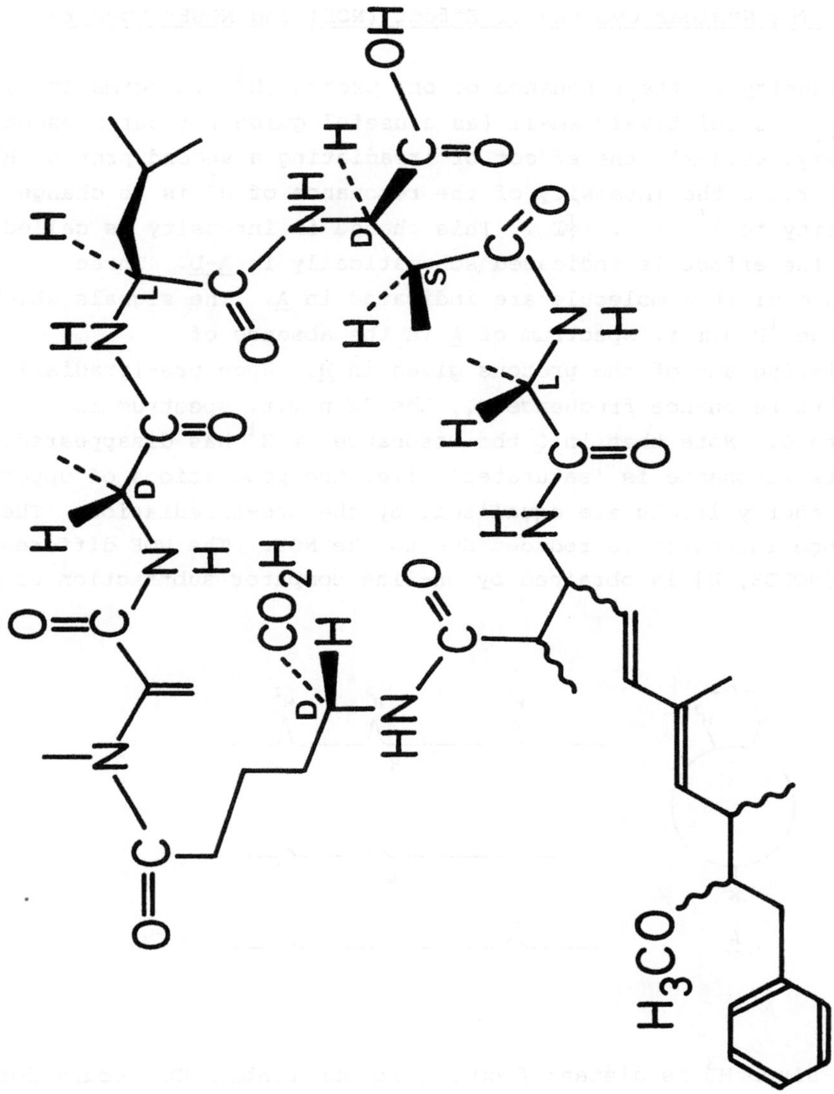

6

Scheme 4 - The Nuclear Overhauser Effect (NOE) and NOSEY Spectra

If the intensity of the resonance of one proton (H^1) is normally I, then if $r_{1,2}$ is relatively small (as a useful guide for our present purposes say, <0.4nm), the effect of irradiating a second proton (H^2) before recording the intensity of the resonance of H^1 is to change its intensity to I', i.e. $I \neq I'$. This change in intensity is called the NOE. The effect is indicated schematically in A-D. Three hydrogen nuclei in a molecule are indicated in A. The signals which arise in the ^1H n.m.r. spectrum of A in the absence of pre-irradiating any of the protons given in B. Upon pre-irradiation of H^1 at its resonance frequence ν_1, the ^1H n.m.r. spectrum is modified to C. Note that in C the resonance of H^1 has disappeared, because its resonance is 'saturated', i.e. the populations of upper and lower energy levels are equalized, by the pre-irradiation. The H^2 resonance intensity is reduced due to the NOE. The NOE difference spectrum [NOEDS, D] is obtained by on-line computer subtraction of C from B.

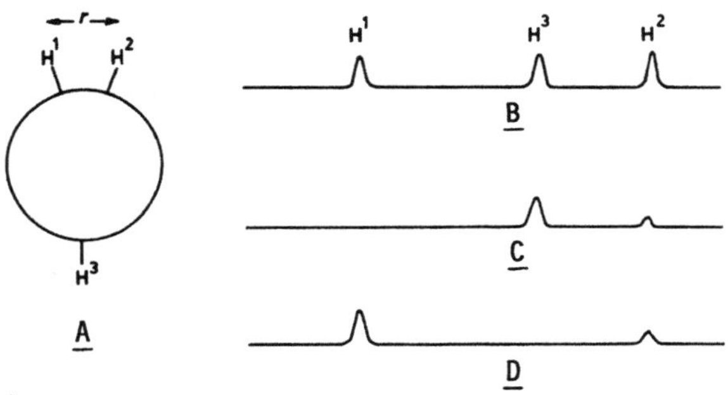

Note that since H^3 is distant from H^1, no observable NOE occurs for H^3. The value of the difference spectrum is evident for molecules containing many protons; the NOEDS then contains only signals due to the irradiated proton H^1, and less intense signals due to those protons which are proximate to H^1. Such experiments are extremely valuable in determining structures of molecules in solution.

A very important variety of the proton nmr spectrum is a NOESY spectrum. This is a two dimensional spectrum in which all the NOES occurring in a molecule, instead of being recorded in a large number of separate NOE difference spectra, are obtained in a single experiment.

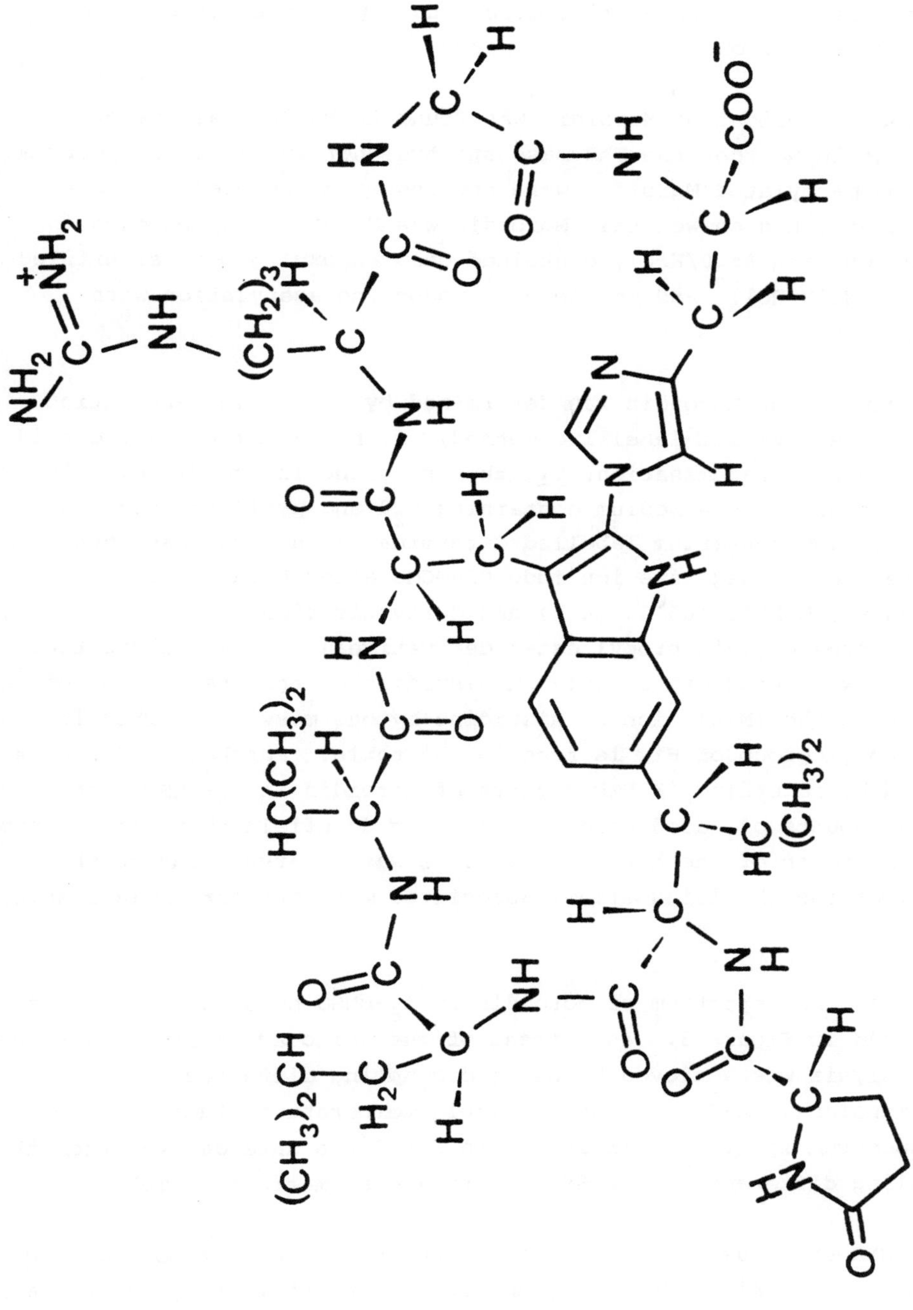

I

imidazole ring has a free imino group and a proton attached at the 2-, 4- or 5-position.

The molecular weight of Moroidin was found to be 986 Daltons by FABMS, but in neither the FAB mass spectrum nor the EI mass spectrum of the permethylated Moroidin were any sequence ions seen. Further FABMS experiments showed that Moroidin was N-terminally blocked (no acetylation with Ac_2O/H_2O), contained one carboxyl group (esterified once with MeOH/HCl), and no alcohol groups (no acetylation with Ac_2O/pyridine).

The C-terminus of Moroidin was determined by selective deuteration (the modified tritium-labelling method) of the C-terminal amino acid residue through racemisation, via the oxazolone intermediate, using acetic anhydride in a medium containing D_2O and pyridine (see also earlier). The deuterium labelled C-terminal amino acid was then characterised by negative ion mode CI-GCMS after total acid hydrolysis (6M HCl, $105^{0}C$, 24 h) and derivatisation (N-trifluoroacetyl-isopropyl ester derivatives). It was found that deuterium was incorporated only in histidine by the observation of an increase in the $(M-H)^-$ ion of histidine by one mass unit. Therefore, the carboxyl group of His is free in the toxin. Further evidence was obtained by recording ^1H NMR spectra of Moroidin in d_6-DMSO with varying amounts of added acid. A limiting downfield shift of 0.2 ppm on the α-proton of the histidine residue was observed, due to the removal of the shielding effect associated with the carboxylate anion of His.

A 500 MHz ^{1} NMR spectrum of Moroidin in d_6-DMSO solution at 333K is reproduced in Figure 3. Assignment of the amino acids found by amino acid analysis was achieved by using decoupling difference, two-dimensional ^1H-^1H COSY and variable temperature studies. The COSY spectrum is also shown in Figure 3 and the data derived from the decoupling difference and COSY spectra are summarised in Table 3.

The ^1H NMR spectrum of Moroidin indicated that the β-protons of the glutamic acid residue were very non-equivalent (δ 1.76 and 2.26 ppm) and their γ-CH_2 signal was very complex compared to the simple triplet expected for a normal glutamic acid residue . This indicated very restricted rotation and so it seemed likely that Moroidin contained a pyroglutamyl residue rather than a normal glutamyl

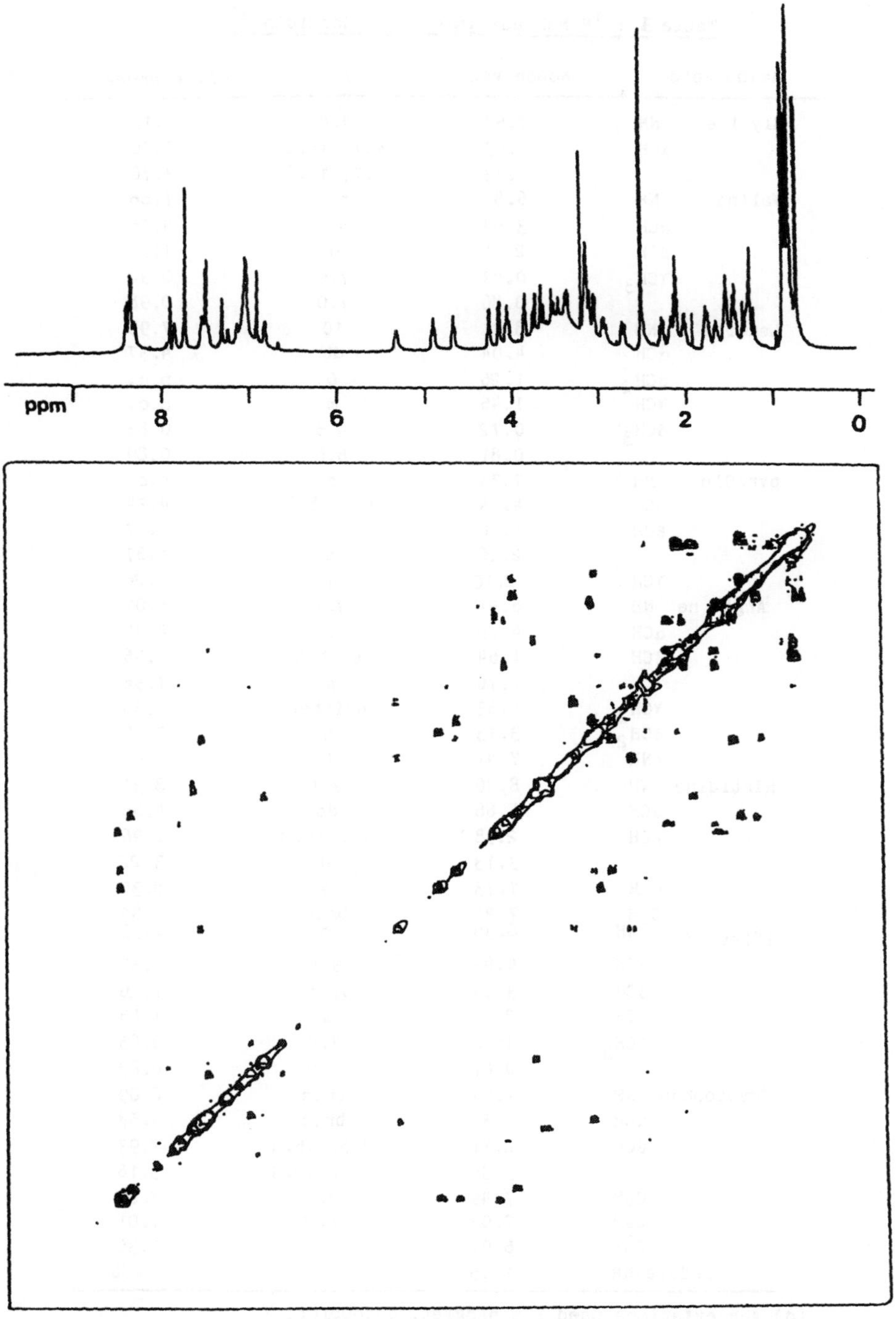

Fig. 3 - 500MHz ^{1}H NMR spectrum of Moroidin (~14mM) in d_6DMSO solution at 333K. The COSY spectrum at 400 MHz is shown below.

Table 3 : ^{1}H NMR assignment for Moroidin[a]

Amino acid		δobserved	J	δliterature
Glycine	NH	7.53	4.8	8.15
	αCH	3.63	4.7, 15.8	3.76
		3.73	4.7, 15.8	3.76
Valine	NH	6.81	8.2	7.88
	αCH	3.84	8.2	4.26
	βCH	2.01	o	1.97
	γCH_3	0.83	7.0	0.92
		0.86	7.0	0.98
Leucine	NH	8.31	10	7.95
	αCH	4.04	dt	4.37
	βCH_2	1.26	O	n.c.
	γCH	1.46	m	n.c.
	δCH_3	0.72	6.5	0.85
		0.81	6.6	0.89
pyroGlu	NH	7.73	s	n.c.
	αCH	4.14	4.1, 8.8	4.34
	βCH	1.76	m	1.87
		2.26	m	1.87
	γCH_2	2.10	m	2.26
Arginine	NH	8.41	7.1	8.07
	αCH	4.25	q	4.38
	βCH	1.64	sextet	1.56
		1.76	m	1.56
	γCH_2	1.53	quintet	1.56
	δCH_2	3.13	O	3.17
	εNH	7.41	O	-
Histidine	NH	8.36	9.1	8.31
	αCH	4.66	dt	4.66
	βCH	2.98	10, 15.3	2.96
		3.13	O	3.22
	C_2H	7.73	s	8.35
	C_4H	7.31	br.s	7.35
β^sLeu[d]	NH	8.36	d	7.95
	αCH	4.89	9.4	4.37
	βCH	3.04	3.5, 11.7	1.26
	γCH	2.16	m	1.46
	δCH_3	0.80	7.0	0.85
		0.87	7.0	0.89
Tryptophan	NH	7.53	br.s	8.09
	αCH	5.32	br.s	4.58
	βCH	2.71	8.5, 15.3	2.93
		3.35	6.4, 15.3	3.16
	C_4H	7.49	8.2	7.65
	C_5H	7.03	8.2	7.01
	C_7H	6.91	s	7.35
	indole NH	11.5	s	10.76

(d) Abbreviations used : s-singlet, d-doublet, dd-doublet of doublets, dt-doublet of triplets, m-multiplet, br.-broad, O-not measured due to overlap.

residue. This conclusion was supported by partially hydrolyzing the Moroidin with 6M HCl at 110^{0}C for 10 min after which addition of water was detected by FABMS. Acetylation (Ac_2O/H_2O) and esterification (MeOH/HCl), followed by determination of the molecular weights of the derivatives by FABMS, showed the product to contain an amino group and two carboxyl groups. Loss of 129 Daltons (i.e. Glu) from the $(M+H)^+$ ion was observed after one cycle of Edman degradation. Furthermore, loss of 111 Daltons was found when the FAB spectrum of Moroidin was re-run using a higher kinetic energy (9.5 KeV) of xenon atoms. The blocked N-terminus was thus found to be an N-terminal pyroglutamic acid residue. However, attempts to cleave or open the pyroglutamic acid residue by pyroglutamyl aminopeptidase (an enzyme which cleaves pyroglutamic acid specifically), or by selective methylamine cleavages, failed.

Partial Sequence of Moroidin

Moroidin was found to be resistant to enzymic cleavage with chymotrypsin, carboxypeptidase Y, thermolysin, papain and Staphylococcus protease. However, it could be opened by trypsin (which cleaves specifically at the C-terminal side of Arg and Lys). Addition of 18 Daltons was then observed by FABMS. Acetylation and esterification showed that this product contained an amino and two carboxyl groups. After one Edman degradation cycle on the tryptic digest of Moroidin, the FAB mass spectrum of the remaining peptide showed that N-terminal glycine has been cleanly removed. The second Edman cycle failed as evidenced by the increase in the molecular weight of the product of 135 Daltons, as determined by FABMS. The increase in mass was due to the coupling of the peptide with PITC. This showed that the next residue is not a standard amino acid, or alternatively that the linkage may be stereochemically hindered so that cyclisation cannot occur.

In order to reveal the nature of the next amino acid residue, the first Edman cycle was repeated on the product of tryptic digestion, and this product was then protected with dansyl chloride (4-N,N-dimethyl-amino-1-naphthalene sulphonyl chloride). Upon total hydrolysis of this material (6M HCl, 105^{0}C, 24 h), the N-terminal residue was identified as histidine by the use of TLC on polyacrylamide plates; the dansylated N-terminal residue showed the same chromatographic properties as the authentic dansyl-histidine.

Since trypsin cleaves specifically at the carboxyl end of arginine, it is obvious that the newly generated C-terminus in the tryptic digest is arginine. Therefore, the partial sequence of Moroidin is:

-Arg-Gly-His(OH)-

It became obvious that the histidine was linked in the intact molecule through its imidazole moiety (note that some unusual linkage of histidine is required, since its carboxyl group has already been shown to be free, and even after liberation of its amino group in the above degradative sequence, it still remains attached to the remainder of the molecule). The negative reaction with Pauly's reagent indicates that the histidine's imidazole ring could be substituted at the NH, C_2 or C_4 position(s).

The ^1H NMR spectrum of Moroidin showed the presence of the C_2 and C_4 protons of histidine (δ 7.73 and 7.35 ppm) by comparison of the observed chemical shifts with the literature values (δ 8.35 and 7.35 ppm), deuterium exchange and pH dependence studies. In addition, no resonance assignable to an imidazole NH was found. It could thus be concluded that the histidine residue of Moroidin is substituted at the imidazole NH.

Nature of the Additional Fragments

FABMS showed that the molecular weight of Moroidin was larger than the sum of the masses of the amino acids found by amino acid analysis. By peak matching, an exact mass of 986.50718 (average value) Daltons was obtained for Moroidin using Substance P (4-11), Dynorphin (1-8), and Oxytocin as references. This determination gave a molecular formula of $C_{47}H_{66}O_{10}N_{14}$ (calculated=986.5086) for Moroidin and indicated that the mass and molecular formula of the additional fragments are 296.1399 and $C_{17}H_{18}O_2N_3$ respectively. The structures of the fragments giving rise to this mass difference were partially revealed by NMR spectroscopy.

The ^1H NMR spectrum of Moroidin appeared to show the presence of two novel amino acids. One of them gave rise not only to an α-proton signal (δ 4.89 ppm) coupled to amide (δ 8.36 ppm) and β-proton (δ 3.04 ppm) resonances, but also a γ-proton (δ 2.16 ppm) coupled to the same β-proton and to two methyls (δ 0.80 and 0.87 ppm). This was assigned to a β-substituted leucine residue (A) in which the substituent (X) could be either an aromatic system or an electron withdrawing group.

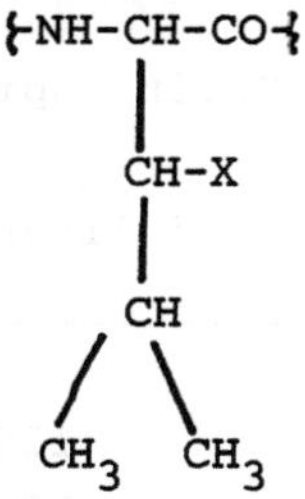

A

The other amino acid was found to be a substituted tryptophan residue. Evidence for this was as follows. First, a λ_{max} at 275 nm was observed in the UV spectrum of Moroidin. Second, the ^{1}H NMR spectrum contained an α-proton signal (δ 5.32 ppm; Table 3), coupled to an amide resonance (δ 7.53 ppm) and two non-equivalent protons (δ 2.98 and 3.35 ppm, similar to those of the histidine residue). The degree of non-equivalence of the β-protons suggested that they were attached to an aromatic system. Third, three aromatic protons were found in addition to those assigned to histidine; two of these were coupled to one another with a coupling constant of 8.5 Hz. Fourth, an exchangeable proton signal at δ 11.5 ppm was found which gave an nOe to the third aromatic singlet at δ 6.91 ppm in the ^{1}H NMR spectrum of Moroidin. Fifth, the ^{13}C NMR spectrum showed Moroidin to contain eight additional aromatic carbons. Three of these were shown to be methine carbons by ^{13}C DEPT experiments (13). Comparison of the literature values of ^{13}C chemical shifts of aromatic amino acids with the observed ^{13}C shifts showed they were consistent with the presence of a substituted tryptophan residue. Agreement for the indole moiety was good (±3 ppm; Table 4) except for the C_2 and C_6 carbons (see Fig 4 for numbering of the indole nucleus) which showed a downfield shift of 9 and 7.7 ppm respectively. This suggested that the tryptophan residue was substituted at these positions.

Substitution at the C_2 position of the proposed tryptophan residue was shown by the negative result obtained with Ehrlich's reagent (p-dimethylaminobenzaldehyde/HCl), which reacts with pyrrole derivatives having one intact CH group. Substitution at the C_6 position was shown by nOe and decoupling difference experiments. The α-proton of the tryptophan residue showed an nOe with one of a pair

Table 4

Comparison of the aromatic ^{13}C NMR signals of Moroidin with literature values for tryptophan

	Chemical Shift (ppm)	
Assignment	δ observed	δ literature*
C-2	135.0	126.0
C-3	105.3	108.4
C-4	121.6	120.3
C-5	123.0	122.9
C-6	127.0	119.3
C-7	121.0	118.7
C-8	137.5	137.3
C-9	128.0	127.5

*Measured in the protected linear tetrapeptide TFA-Gly-Gly-X-Ala-OCH_3 (in D_2O solution)

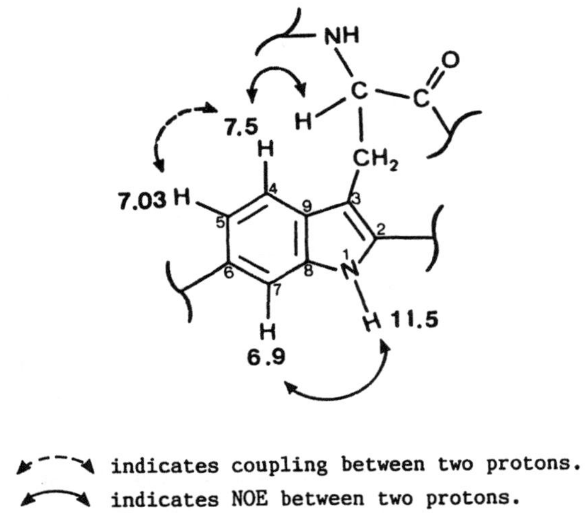

Fig. 4 - Structure of the 2,6-disubstituted trytophan residue deduced from UV, NMR, and chemical evidence

of aromatic doublets (J_{HH}=8.5Hz). This together with the nOe seen between the singlet at δ 6.9 ppm and the NH at δ 11.5 ppm confirmed the substitution at the C_6 position. The results are summarised in Figure 4 with coupling and nOes between two protons indicated by (⇢) and (↷) respectively.

The presence of the β-substituted leucine and 2,6-disubstituted tryptophan residues accounted for the measured molecular weight and formula of Moroidin exactly. Although attempts to isolate the two novel amino acids by alkaline (4M NaOH, 106^{0}, 24 h) and p-toluenesulphonic acid hydrolysis failed, we consider evidence for their presence using ^{1}H and ^{13}C NMR to be satisfactory.

Determination of the Full Amino Acid Sequence using nOe Experiments

Taking the chemical information in conjunction with the amino acid composition, it was clear that Moroidin must be bicyclic, with a partial sequence of -Arg-Gly-His(OH)-. However, no further sequence information could be obtained from either selective enzymic or acid cleavages. Therefore, classical methods or mass spectrometric studies could not readily deal with this problem. In solving this problem, both one-dimensional nOe difference and two-dimensional nOe spectroscopy (NOESY), which give information on relative interproton distances, were used. All nOe experiments were done at room temperature (293K) using d_6-DMSO as solvent for the sample. The study of nOes was concentrated primarily on the NH and α-protons to deduce the sequence of the amino acids. The data are tabulated in Table 5 and are summarised in Figure 5. The nOes between two protons are indicated by the arrows (↷).

The observed nOes were shown to be direct (that is not arising via spin diffusion) by studying their rate of growth. This was done by measuring the % nOe growth as a function of the preirradiation time (0.2, 0.4, 0.8, 1.6, 3.2, 6.4 s). This showed that the time dependences of the observed nOes were exponential with characteristic t 's (the time taken for the intensity changes to reach half their maximal values). The t 's are proportional to r^{-6}, where the values of r are the various interproton distances. These allowed us to estimate the relative distances between protons. This gave a sequence of:

Table 5

Nuclear Overhauser effects observed in Moroidin

Proton	NOE observed
Trp C_4	Trp C_5, Trp αCH
Trp C_5	Trp C_4, β^sLeu αCH
β^sLeu αCH	Trp C_5, Leu NH
Leu NH	β^sLeu αCH, Val NH
Trp αCH	Trp C_4, Arg NH[(a)], Trp βCH
Trp NH	Val αCH
Gly NH	Gly αCH, Arg αCH
Gly αCH	Gly NH, His NH[(a)]
pyroGlu αCH	β^sLeu NH[(a)]
β^sLeu βCH	Trp C_7, β^sLeu NH[(a)]
His αCH	His C_4
Trp NH_{indole}	Trp C_7

[a]Revealed after acid titration.

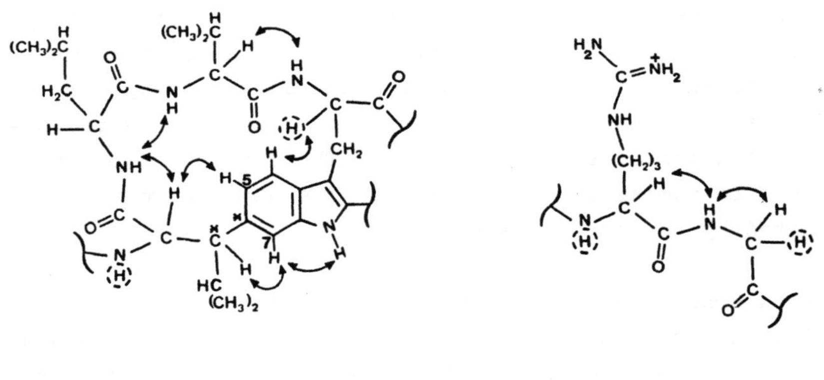

Fig. 5 - Partial amino acid sequences of Moroidin derived from proton NOES (indicated by ↔). The sequence Arg-Gly is independently available from enzymic digestion and mass spectrometric experiments.

$-\beta^{S}$Leu-Leu-Val-Trp-

(where β^{S}Leu represents the β-substituted leucine unit) which was then linked to form a cyclic structure through a C*-C* bond (Figure 5). Evidence for this linkage was the observation of fast growing nOes given by the C_5 and C_7 protons of the tryptophan residue to the α- and β-protons of the β-substituted leucine, respectively. The structural conclusion derived in the left hand portion of Figure 5 is supported by the subsequent demonstration (see later) that the NH of the β-substituted leucine residue is attached to the available carbonyl group of pyroglutamic acid.

The sequence of -Arg-Gly- had already been found by tryptic digestion and mass spectrometric studies as discussed above. In addition, nOes were also observed for the protons encircled (Figure 5) to the NH protons at δ 8.2-8.3 ppm. These amide protons (β^{S}Leu, Arg and His) overlapped with one another and so linkages were ambiguous at this stage.

The connections of all the amino acids were more clearly established by recording a combined COSY-NOESY spectrum of Moroidin under acidic conditions so that the above mentioned amide protons were well-separated. The derived connections are shown in Figure 6 and are represented by the symbol •—•.

An important feature of Moroidin not yet discussed is the C-N linkage between the histidine and tryptophan residues (Figure 6) which completed the tricyclic structure of Moroidin. This was suggested by the nOe observed beteen the C_2 proton of the histidine residue and the indole NH of the tryptophan residue. The other N atom of the imidazole may equally be attached to C-2 of the indole.

Oxidation of Moroidin

The C-N linkage was demonstrated chemically by oxidising Moroidin with bis-trifluoroacetoxy-iodobenzene (TIB). TIB is normally used for Hofmann degradation to convert an amide function to an amine. It contains trivalent iodine which makes it a potential oxidizing agent (see also earlier). Moroidin was allowed to react with TIB in CH_3CN/H20 (1:1) at room temperature in the dark. The product was

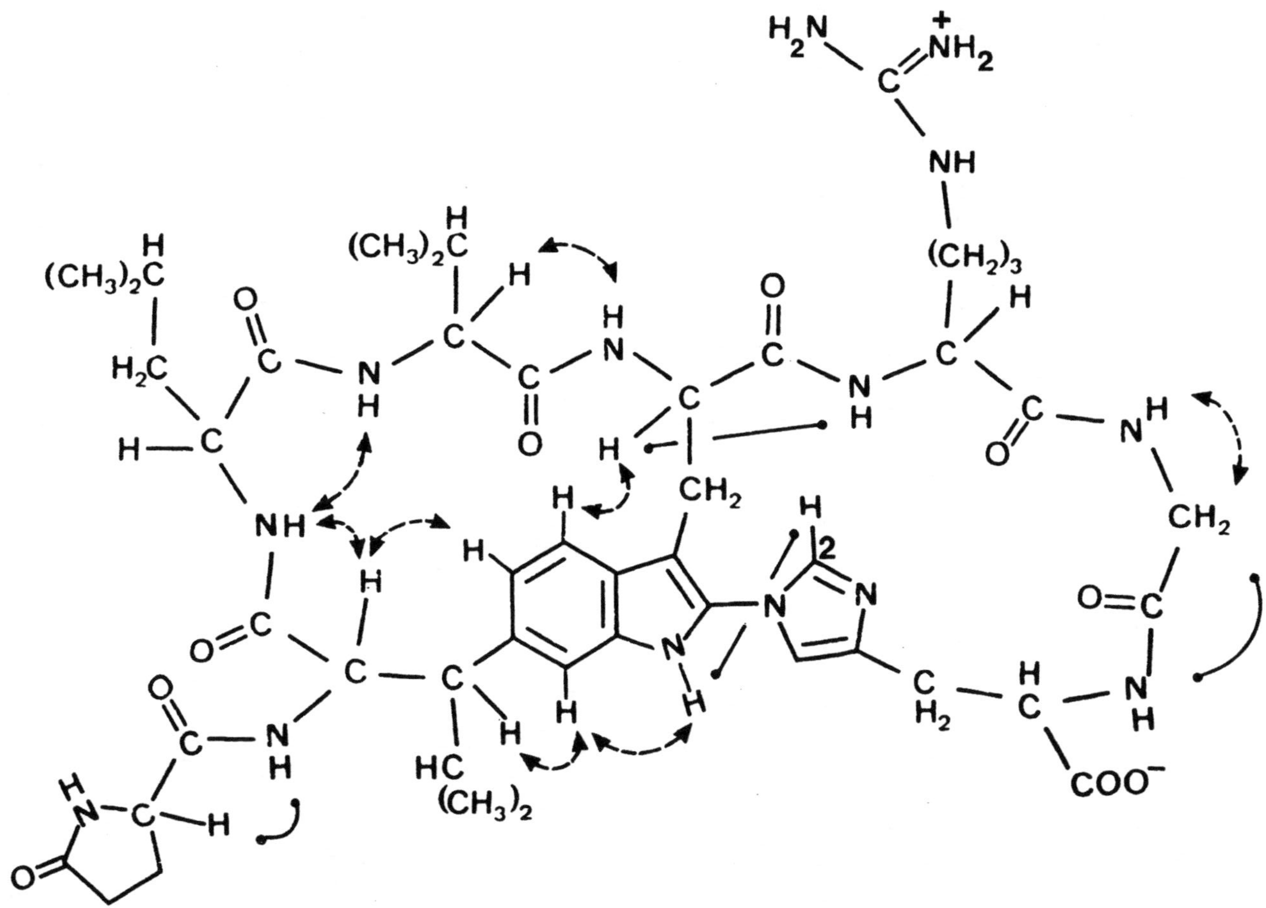

Fig. 6

analysed by FABMS. FABMS showed that the product was a mixture of components with molecular weights of 1054 and 1020 daltons. The mixture was the analysed by HPLC and one major product (OPI) was isolated and characterised by FABMS, UV and ^1H NMR spectroscopy.

FABMS showed the molecular weight of OPI to be 1020 daltons. Accurate mass measurement gave an exact mass of 1020.5151 and a molecular formula of $C_{47}H_{68}O_{12}N_{14}$ (calculated=1020.5142) using [Val^5]-Angiotensin II (M_R=1031) and Oxytocin (M_R=1006) as references. This indicated that the extra mass of OPI compared to Moroidin was due to the addition of two hydrogen and two oxygen atoms. Acetylation (Ac_2O/H_2O) and esterification (MeOH/HCl) showed the oxidation product contained no additional amino or carboxyl groups.

In the positive ion mode FAB mass spectrum, sequence ions which appeared as doublets separated by 16 mass units were seen at m/z 883, 826 and 670 from the pseudomolecular ion of OPI (Figure 7). Consideration of the mass difference between the sequence ions and comparison of these differences with those arising from the peptide fragmentations commonly occurring in FAB mass spectra gave the sequence of -Arg-Gly-His-OH. Doublets separated by 16 mass units are observed (see Fig. 7); their appearance may be due to the oxidised nature of OPI. The C-terminal histidine was confirmed by performing a carboxypeptidase Y digestion. Loss of 137 daltons was observed in the FAB mass spectrum after the digestion. This suggested that the imidazole NH of histidine which was absent in the original Moroidin was generated during the reaction. This was further supported by the observation of an NH resonance at δ 18.4 ppm in the ^1H NMR spectrum of OPI. Furthermore, a positive reaction with Pauly's reagent was observed with OPI.

When the oxidation reaction of Moroidin was repeated in a deuterated medium (CD_3CN/D_2O), it was found that no deuterium was incorporated as no change in molecular weight was observed by FABMS. This implied that the two added hydrogens in OPI are exchangeable and are not carbon-bound protons.

In the ^1H NMR spectrum of OPI (Figure 8) all the resonances of Moroidin were found. Most shifts agreed to within 0.3-0.5 ppm, except those of the signals due to the amide protons and the β-protons of the tryptophan residue. Any newly generated protons and

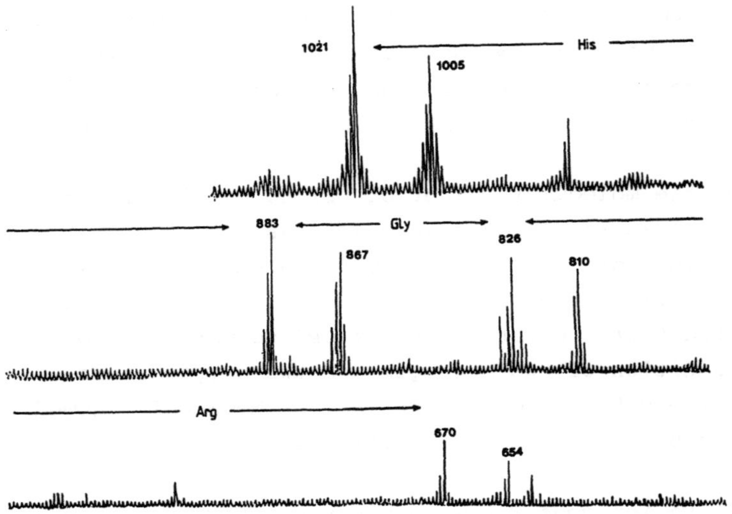

Fig. 7 - High mass region of the positive ion FAB mass spectrum of OPI, obtained from Moroidin by oxidation with TIB.

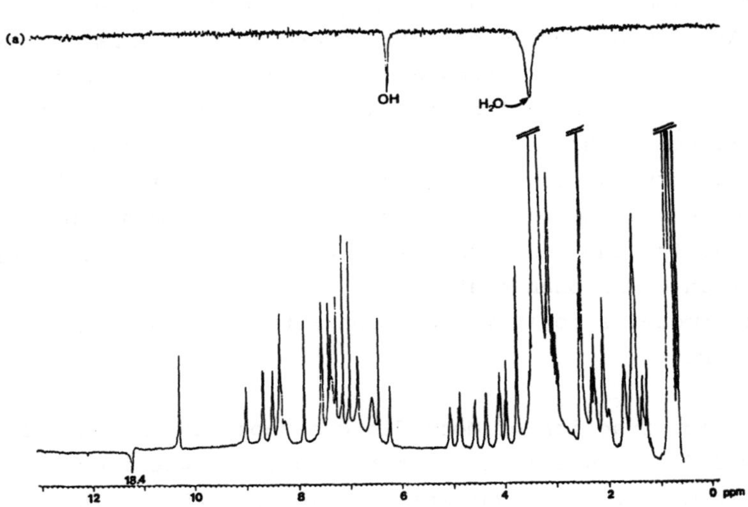

Fig. 8 - ^{1}H NMR spectrum of OPI in d_6-DMSO. The superimposed trace (a) is a difference spectrum derived by subtraction of the spectrum obtained upon irradiation of the water peak from the normal spectrum.

those that suffered a relative chemical shift of > 0.15 ppm or more on the oxidation of Moroidin to OPI are shown with the shift (in ppm) adjacent in Figure 9, where no stereochemical detail is given.

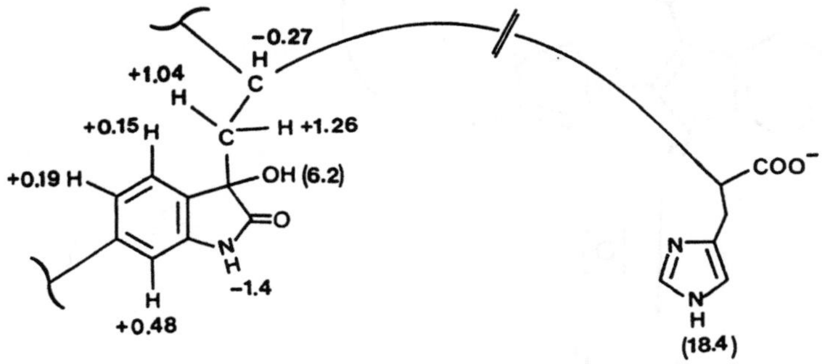

Fig. 9 - Modified structural units deduced to be present in OPI. Chemical shifs of newly generated protons are indicated in parentheses. Changes in chemical shift > 0.15 ppm are also indicated; a "+" sign indicates an upfield chemical shift.

The β-protons of the tryptophan residue showed a significant upfield shift (i.e. from δ 2.71 and 3.35 ppm to δ 1.45 to 2.29 ppm) suggesting that part of the aromatic system of the indole moiety had been destroyed. This was further supported by the UV absorption of OPI at 269 nm (compared with 275 nm for the intact Moroidin). The hypsochromic shift of the λ_{max} is consistent with the conversion of the indole to an oxindole. The presence of a hydroxyl group was revealed by a ^{1}H NMR saturation transfer difference experiment performed in d_6-DMSO. The difference spectrum is shown in Figure 8(a). The water signal, which was saturated in the experimental data set, appears as a negative peak, as does the hydroxyl proton which has become saturated, or partially saturated, by chemical exchange with water in the solvent.

Attempts to acetylate the hydroxyl group of OPI with either acetic anhydride/pyridine or acetic anhydride/dimethylaminopyridine failed (the latter probably due to solubility problems). Since the hydroxyl group is not coupled to any protons as observed by decoupling difference experiments, we conclude that it is tertiary. Taking all this information, it became clear that the oxidation product, OPI, of Moroidin was (B), which is consistent with the mechanistic scheme shown in Figure 10. It is proposed that reaction is occurring

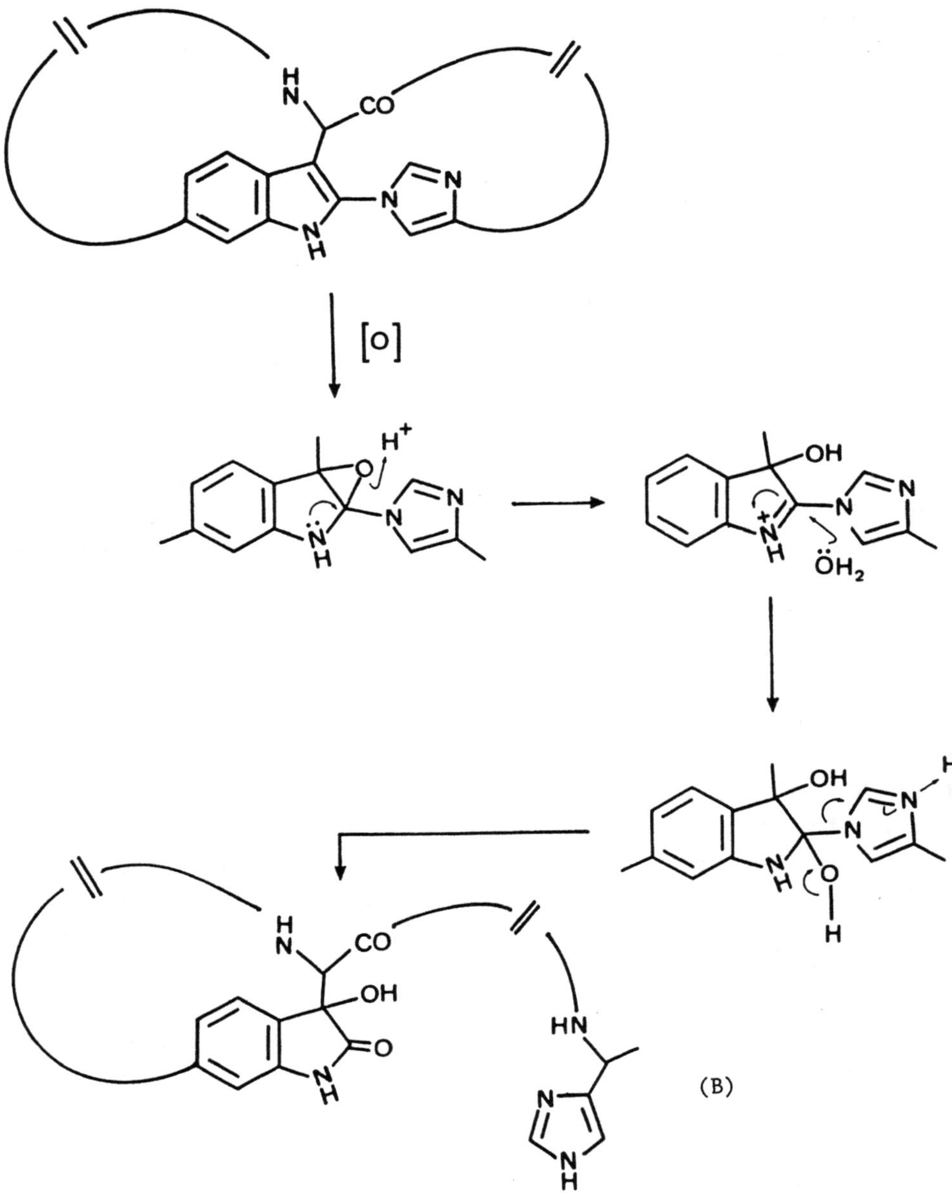

(B)

Fig. 10

through oxidation of the indole ring, followed by a nucleophilic attack of water, to give the product OPI.

Stereochemistry of the Amino Acids

From the enzymic digestion of Moroidin by trypsin and the carboxypeptidase Y cleavage of OPI, we knew that arginine and histidine must possess the L absolute configuration. The configurations of the other amino acids, except glycine (which is not chiral), β-substituted leucine and tryptophan, were determined by GC analysis on a capillary column coated with a chiral stationary phase (Chirasil-Val). Moroidin was hydrolysed into its amino acid components, and these were derivatised to their N-trifluoroacetyl-isopropyl esters (isopropanol/HCl;$(CF_3CO)_2O/CH_2Cl_2$), and detected, after passage through the GC column, by flame ionization. Co-injection with derivatives of authentic D- and L-amino acids showed that the configurations of pyroglutamic acid, leucine and valine are L.

The stereochemistries of the β-substituted leucine and tryptophan residues have not yet been determined since these amino acids could not be isolated. Structure 7 shows the stereochemistry of 6 of the 8 amino acid residues of Moroidin.

(iii) Structure Elucidation of the Vancomycin Group of Glycopeptide Antibiotics, and Determination of their Mode of Action

Single cell organisms go to a great deal of trouble to synthesise unusual peptides. Since these organisms may typically utilise some tens of kilo-bases of DNA to direct the synthesis of such peptides, it appears likely that the peptides, many of which have antibiotic properties, are produced because the producing organism is thereby given some selectional advantage in its fight for survival. Among such peptide antibiotics are those of the vancomycin group.

Vancomycin is a glycopeptide antibiotic obtained from *Streptomyces orientalis*, and it is bactericidal against Gram-positive bacteria. It was first isolated in 1956(14), and now enjoys widespread clinical use, especially in the treatment of severe staphylococcal infections (e.g., endocarditis and wound septicaemia). A recent development has been its use orally in the treatment of pseudomembranous colitis, a severe and potentially lethal disease usually associated with antibiotic treatment after major gastrointestinal surgery (15).

In the 1960s, methods suitable for the structure elucidation of vancomycin were not available, and early workers had to be content with the isolation of several important degradation products which retained the vast majority of the structure and with the characterization of N-methylleucine, aspartic acid, glucose, and chlorophenols, obtained as degradation products, each of which accounted for a small part of the structure (16,17). However, with the advent of proton nuclear magnetic resonance at 270 mHz in the 1970s, it was possible to assign fully the ^{1}H NMR spectrum of vancomycin (18) in terms of all the structural units which had been isolated via chemical degradation. Moreover, it was possible to assemble these units into a partial structure, largely by taking advantage of negative nuclear Overhauser effect (nOe). It is interesting to note that the discovery of these negative nOes (for vancomycin in Me_2SO-d_6) was purely serendipitous. In the process of assigning the spectrum by spin decoupling, it was noted that other resonances were reproducibly reduced in intensity. It must be remembered that difference spectroscopy was not routinely available at this time, and hence positive nOes, which are normally small (1-5%), had not proved to be of wide utility and reliability. However, the negative nOes observed for vancomycin in Me_2SO-d_6 were large (15-50%), and hence readily observed without the use of difference spectroscopy.

The nOes are negative because vancomycin in Me_2SO-d_6 has a relatively slow tumbling time (ca.10^{-8}s). This is due to a combination of its relatively large molecular weight (1448) and the use of a relatively viscous solvent. Despite the establishment of a partial structure, there were too many asymmetric centers in the molecular to permit the determination of a complete structure. However, the early nOe work was crucial in showing the power of a method which has been used in establishing the structures of all other members of the vancomycin group.

The next important step for structure determination of the vancomycin group was the determination of the X-ray structure (Fig.11) of a degradation product of vancomycin, CDP-I (19). The conversion to CDP-I is carried out by heating vancomycin at 80-90 ^{0}C in water at pH

4.2 for ca. 2 days. In this conversion a primary amide of vancomycin is converted to a carboxyl group. Since earlier NMR studies (18) had established that the carboxyl group of the C-terminal meta-hydroxylated phenylglycine was free, it was assumed that in vancomycin, the carboxyl group of the iso-Asp residue of CDP-I was present as a primary amide, i.e., that vancomycin contains an isoasparagine residue and is otherwise the same as CDP-I.

The first evidence that the above assumption was incorrect came from further NOE studies carried out on vancomycin (20). It might be asked why the further nOe studies were being carried out when it was assumed that the structure of vancomycin was known. The reason was that we had posed the hypothetical question: "Now that nOe difference spectroscopy was available, would it have been possible to elucidate the structure of vancomycin, with stereochemical detail, from ^{1}H NMR data without the help of X-rays?" Since the interpretation of the data obtained might at some point have been aided unconsciously by hindsight, it is even now difficult to answer the question unequivocally. However, two important points were clear. First, certainly a great deal of the structure could be put together with complete stereochemical detail in agreement with that found in the X-ray of CDP-I. Second, nOes showed that the chlorine atom attached to the ring labeled 2 in Figure 11 is in fact at the back of the molecule (see double-headed arrow in Figure 11) in vancomycin. Thus, this ring has rotated by ca. 180^0 in going from vancomycin to the crystalline CDP-I. If this were an equilibrium process, it might be regarded as a process of little significance. However, CPK models suggested that rotation of ring 2 would have been impossible at the temperatures used for the conversion of vancomycin to CDP-I. It was therefore concluded that bond fission (and re-formation) in the 17-membered macrocycle containing ring 2 (Figure 11) had occurred during CDP-I formation. This could be accommodated by a reversible retro-aldol reaction which serves to break and remake the bond indicated by an arrow in Figure 11. However, Harris and Harris noted that it is not clear why such a process should occur with retention of the stereochemistry at the carbon atoms at the site of the proposed bond cleavage. They carried out experiments which showed (21) that vancomycin (Figure 12) contains an asparagine rather than an isoasparagine residue. At pH 4.2, asparagine isomerizes to isoaspartic acid via a substituted succinimide intermediate. Whereas the chlorine-containing ring 2 is indeed unable to rotate in

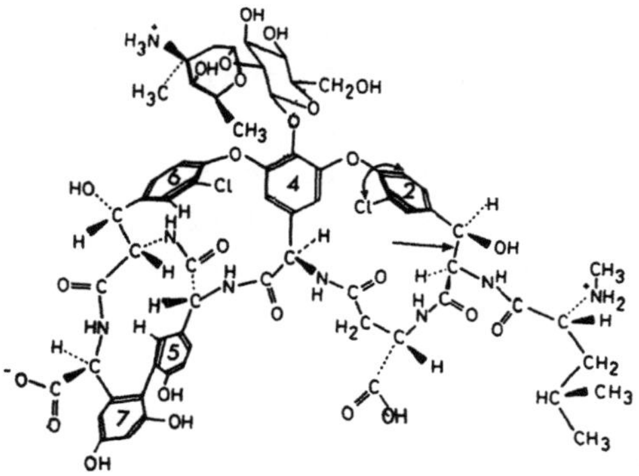

Fig. 11

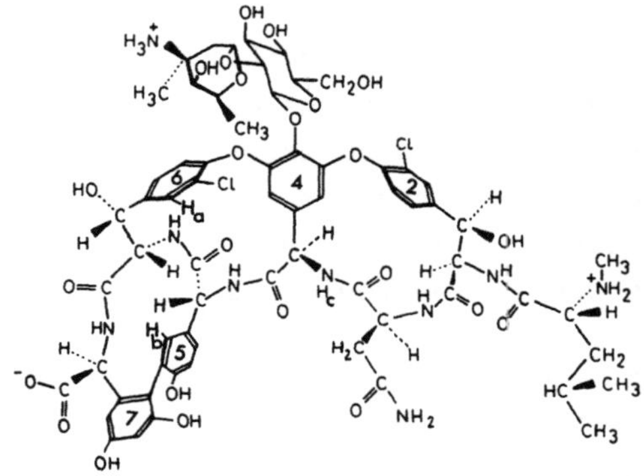

Fig. 12

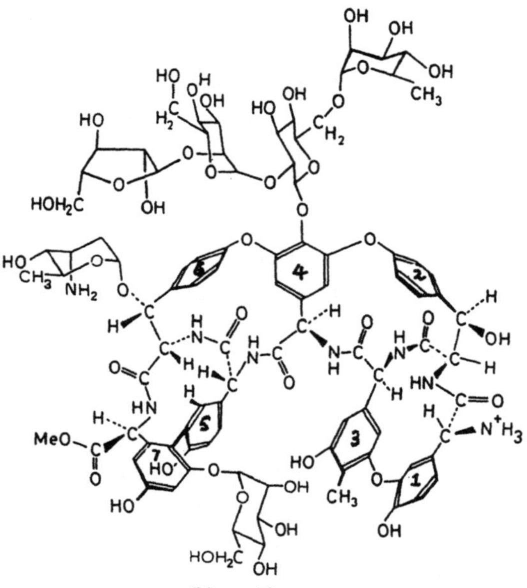

Fig. 13

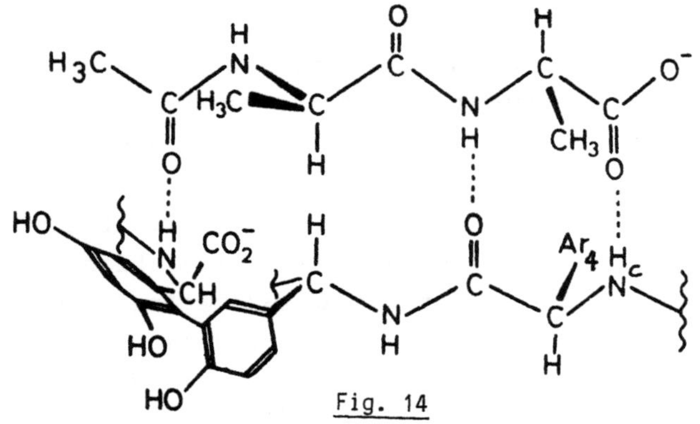

Fig. 14

vancomycin (Figure 12, containing the 16-membered macrocycle), it can do so slowly in CDP-I (Figure 11, containing the 17-membered macrocycle). Note that in CPK models both these rotations are utterly impossible, and in this respect the CPK models are misleading.

The early ^{1}H NMR work and X-ray study on vancomycin and its derivatives led to a situation where, using chemical and ^{1}H NMR methods alone, it has been possible to elucidate the structures of several other members of the vancomycin group. These include ristocetin (22,23), avoparcin (24), antibiotic A35512 (25), actaplanin (A-4696) (26), and teicoplanin (27). All the antibiotics are linear heptapeptides (cf. Figure 12), and the unusual amino acid residues are linked together by several phenol oxidative couplings. The structure of ristocetin A, in the determination of which chemical experiments were heavily complemented by ^{1}H nOes, is reprodced in Figure 13. All the compounds share identical oxygen substitution patterns on rings 4,5,6 and 7. The stereochemistry at the asymmetric centers associated with the seven amino acids is conserved, and from the N-terminus is R,R,S,R,R,S,S.

Biosynthetic studies have been carried out on vancomycin (28) and ristocetin A (29). It was found that both of the β-hydroxychlorotyrosine units of vancomycin and both the β-hydroxytyrosine units of ristocetin A can be derived from tyrosine. It was also established that the p-hydroxyphenylglycine units (numbering amino acid residues 1-7 from the N- to the C-terminus) at positions 4 and 5 in vancomycin and at positions 1,4, and 5 in ristocetin A can be derived from tyrosine. The m-dihydroxylated phenylglycine units (residue 7 in vancomycin, and residues 3 and 7 in ristocetin A) are derived from acetate. The aromatic methyl group of residue 3 in ristocetin A is presumably derived from S-adenosylmethionine, but this has not be established experimentally. The polycyclic structures are formally completed from a heptapeptide by three phenol oxidative couplings (two C—O and one C—C) in vancomycin and four (an additional C—O) in the case of ristocetin A, and by the addition of sugars. However, the order of these steps has not been established.

<u>Molecular Basis of Action of the Vancomycin Group</u>

With the above structures established, it was possible to probe the molecular basis of action of the antibiotics. The vancomycin group of antibiotics are inhibitors of bacterial cell-wall biosynthesis. A key finding was that vancomycin forms strong complexes with the cell-wall peptidoglycan precursor UDP-N-acetylmuramyl-pentapeptide (30). By stepwise degradation of such a precursor, Perkins (30) was able to show that an acyl-D-alanyl-D-alanine C-terminus is the feature essential for complex formation. The molecular basis for complex formation was therefore probed by comparing the proton NMR spectra of the cell-wall analogue N-Ac-D-Ala-D-Ala/antibiotic complexes. The first study in this area (31) was carried out with the complex N-Ac-D-Ala-D-Ala/vancomycin before the structure of the antibiotic was established. Although few specific conclusions could be reached therefore, this study was important in establishing the idea of using ^1H NMR in this way and in showing that the C-terminal alanine methyl resonance of N-Ac-D-Ala-D-Ala is shifted to high field by ca. 0.9 ppm upon complex formation. Thus it was concluded that this methyl group lay over the face of a benzene ring of the antibiotic in the complex.

The later ^1H NMR work(18) on the structure of vancomycin had also reported details of the ^1H spectra of an equimolar mixture of vancomycin and N-Ac-D-Ala-D-Ala. Since the 1:1 complex was in fast exchange with the free components at room temperature in Me_2SO-d_6 solution, it was possible to follow the shifts of a number of the antibiotic protons upon complex formation. In particular, protons H_b and H_c underwent marked downfield shifts, and H_a underwent a marked upfield shift (Figure 12). It was therefore concluded that these three protons, which are in a cleft on one side of the antibiotic, are near to the peptide fragment in the antibiotic-peptide complex and that NH_c is hydrogen bonded to a peptide carbonyl group. Furthermore, this carbonyl group is likely to be that of the carboxyl terminus of the peptide since, when this interacts with NH_c, the methyl group of the C-terminal D-alanine can be over the face of benzene ring 4 (Figure 14), thus accounting for its upfield shift When these interactions are optimized, the models indicate clearly that, simultaneously, the two additional hydrogen bonds shown in Figure 14 can be made.

These interactions also cause the secondary methyl group of the N-terminal D-alanine to be over the benzene rings of the biphenyl moiety to give a favorable hydrophobic interaction. These proposals, based on the X-ray structure of CDP-1, have stood the test of time and are all retained in currently accepted models of the binding. We shall see later, however, that they are complemented by further favorable interactions. These further interactions could only be deduced after it was subsequently shown that a portion of the vancomycin molecule, when bound to the cell-wall analogue, is an a conformation different to that found in the X-ray of CDP-I.

Before this conformational change was uncovered, a more complete binding site of ristocetin A for N-Ac-D-Ala-D-Ala was deduced(32). This was possible because, at room temperature in Me_2SO-d_6 solution, the ristocetin A/N-Ac-D-Ala-D-Ala complex is in slow exchange with the free components on the ^1H NMR time scale. Therefore, it was possible, using techniques such as saturation transfer, to correlate the chemical shifts of many of the protons on passing from the free to the bound state. In particular, it was possible to do this for all the NH protons of the secondary amides in both components. This is important since all those NH protons which suffer large downfield shifts upon complex formation are clearly involved in hydrogen-bond formation. Additionally, it was possible to determine the temperature dependence of the chemical shifts of these same NH protons before and after binding. Using the criterion that those amide NHs that show a large temperature dependence (6-10) x 10^{-3} ppm/^{0}C to high field with increasing temperature) are solvent exposed, whereas those that show a low temperature dependence (0-3 x 10^{-3} ppm/^{0}C) are partially or totally hidden from solvent, much about the mode of binding could be deduced.

First, the temperature dependence studies showed that the NH of the C-terminal D-alanine is hidden from solvent after the N-Ac-D-Ala-D-Ala binds, but that of the N-terminal D-alanine is still exposed after binding. Second, the data for the six amide NH's of ristocetin A divide them clearly into two groups. One group (two NH protons) is exposed to solvent before and after binding, and these two protons are on the "back" (as conventionally displayed), convex face of the antibiotic. The second group, of four NH protons, is located at the "front" of the molecule, is located in a cleft before binding (the NH's are partially hidden from solvent), and is clearly involved in hydrogen-bond formation in complexation and/or is even more hidden from solvent after the complex is formed than before.

These experiments established which side of the cell-wall analogue binds to which side of the antibiotic and give, with some help from the earlier work on the vancomycin mode of action, a picture for the molecular basis for complex formation (Figure 15). The main features of the binding are (i) a carboxylate binding pocket in ristocetin A, consisting of NH's 2,3 and 4, which are in a cleft and flanked by hydrophobic walls constituted from the aromatic rings of residues 1 and 2, and the carboxyl group of N-Ac-D-Ala-D-Ala binds into this pocket by hydrogen-bond formation between its oxygen atoms and the NH groups 2, 3, and 4. Note that these hydrogen bonds, which are esentially formed due to electrostatic attractions [$>C{=}O^{\delta-}$---- $^{\delta+}HN<$], should be strengthened in a hydrophobic pocket which provides an environment of low dielectric constant. (ii) The binding of the carboxylate anion into its pocket is slightly strengthened by the favorable Coulombic attraction between the $-CO_2^-$ of the cell-wall analogue and the $-NH_3^+$ at the N-terminus of the antibiotic. (iii) The NH of the C-terminal D-alanine of the cell-wall analogue forms a hydrogen bond to the carbonyl group of residue 4 of the antibiotic; this carbonyl group can be seen at the center of the antibiotic structure shown in Figure 15. (iv) The NH-7 of the antibiotic forms a hydrogen bond to the carbonyl oxygen of the N-acetyl group of the cell-wall analogue. (v) When the above five hydrogen bonds are made, the methyl group of the C-terminal alanine (hidden in Figure 15) forms a hydrophobic bond into the face of the aromatic ring of residue 4 of the antibiotic (this is the benzene ring whose protons are marked by a dot in Figure 15) and the methyl group of the other alanine residue forms a hydrophobic bond over the benzene rings of residues 5 and 7, which constitute the biphenyl moiety.

Analysis of the Molecular Basis for the Antibiotic/Cell Wall Recognition

Given the above information, we have been able to explore some of the factors involved in defining the desired antibiotic geometries, and factors involved in more subtle aspects of binding. This section is concerned with the most recent aspects of this work, and with considerations of how the derived principles can further our understanding of peptide/peptide interactions and protein folding. Interactions of the antibiotics with the above-mentioned dipeptide cell wall analogue, and with N,N-diacetyl-Lys-D-Ala-D-Ala ("tripeptide", an extended cell wall analogue) are considered.

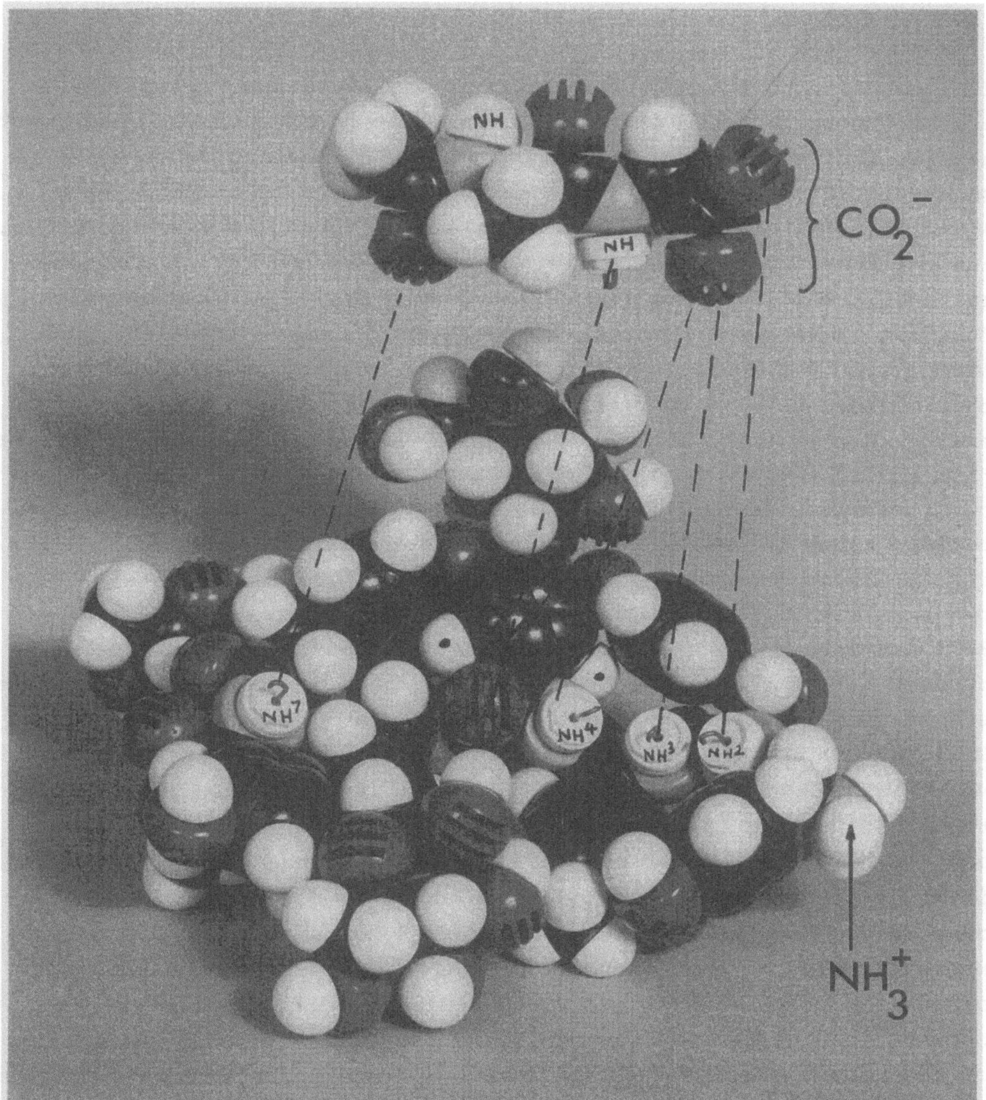

Fig. 15

(a) The Generation of the Carboxylate Anion Binding Pocket

All members of the vancomycin group of antibiotics so far discovered, exhibit in their bound state, a carboxylate binding pocket whose most general feature is three inward pointing NHs, derived from three successive amides of a peptide backbone (the amides of amino acids 2, 3 and 4 from the N-terminus; see Fig. 12). The nature of this pocket, and the three hydrogen bonds formed between it and the carboxylate anion of the cell wall analogue is seen in Fig. 15. A key question is "how are these three NHs made to point in the same direction, when normally in a poly-R or poly-S-peptide in its β-pleated sheet form they would alternate up/down/up ?" We have concluded that the low energy requirement for pocket formation in vancomycin mainly derives from the fact that the amino acids have the R, R, S, and R configurations at positions 1, 2, 3, and 4 from the N-terminus, i.e. residue 3 has the opposite absolute configuration to residues 1, 2, and 4. As a consequence, a β-pleated sheet form would contain the two repulsive 1,3-interactions indicated by double headed arrows in 8. These interactions can be relieved by rotating the backbone to produce 9, but the carbonyl oxygen atom which is omitted in 9 now has an unfavorable steric interaction with the NH group of amino acid 4. This interaction can in turn be relieved by rotating the amide bond which is bounded by arrows in 9 through @ 180°. The conformation 10 results, and this is essentially the shape found in the carboxylate anion binding pockets of the antibiotics. Note that in 10, the three inwardly pointing NH groups of amino acids 2, 3, and 4 can be constrained together if the sidechains of residues 1 and 3 are held approximately coplanar on the top side of 10 as presented, and the sidechains of residues 2 and 4 are similarly held approximately coplanar on the bottom side of 10. In fact, in all known antibiotics of the vancomycin group residues 2 and 4 are indeed constrained near the same plane by covalent cross-linking of their sidechains (see, for example, Figs. 12 and 13). In ristocetin A (Fig. 13) the sidechains of residues 1 and 3 are also covalently cross-linked. The result of the two sets of cross-links makes it appear, from CPK models, that the geometry of the desired binding pocket is constrained by a kinetic barrier. In contrast, in vancomycin, the sidechains of residues 1 and 3 are not covalently cross-linked. Molecular models suggest that in this case, the binding pocket might

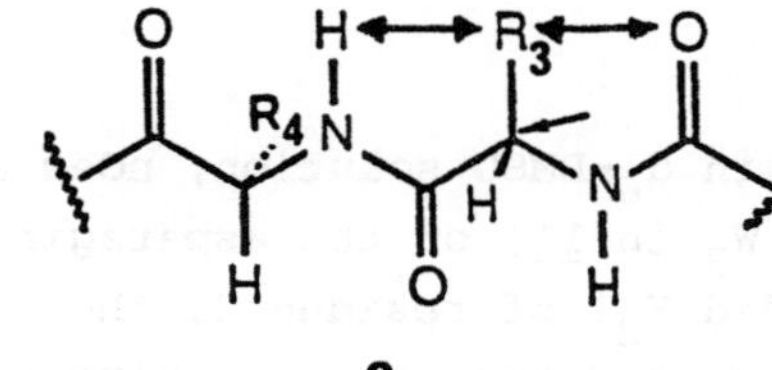

8

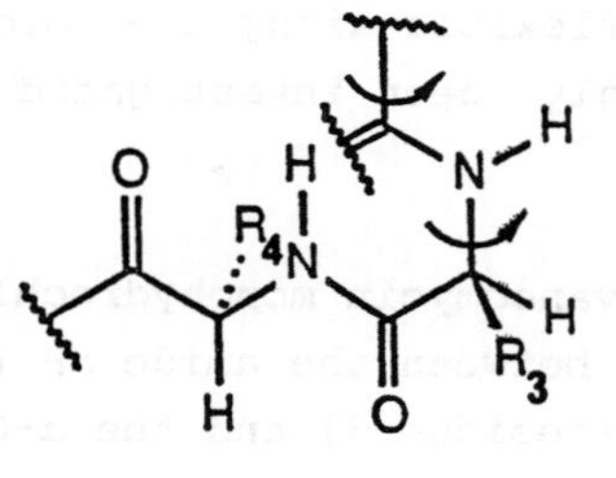

9

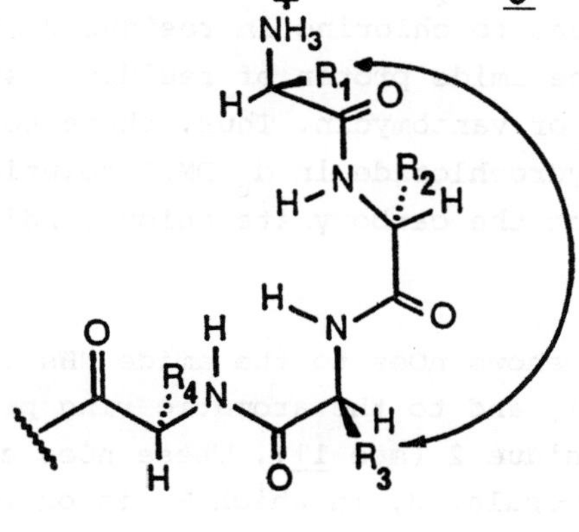

10

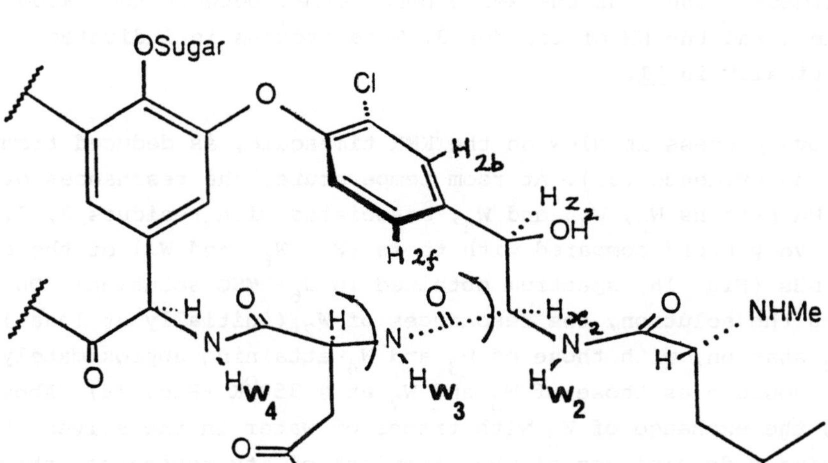

11

be more flexible along the amide backbone, and this possibility has accordingly been investigated by proton NMR spectroscopy.

In free vancomycin monohydrochloride in d_6-DMSO solution, nOes are observed between the amide NH (coded W_3 in 11) of the asparagine residue (residue 3) and the α-CH (coded X_2) of residue 2, the benzylic C<u>H</u>-OH (coded Z_2) of residue 2, and the aromatic ring proton (coded 2b) adjacent to chlorine in residue 2 (see 11). These nOes establish that the amide proton of residue 3 spends some of its time on the back face of vancomycin. Thus, these nOes show that in free vancomycin monohydrochloride in d_6-DMSO solution, a conformation is populated in which the carboxylate anion binding pocket is not present.

However, W_3 also shows nOes to the amide NHs of residues 2 and 4 (coded W_2 and W_4), and to the aromatic ring proton (coded 2f) on the front face of residue 2 (see 11). These nOes establish that a second conformation is populated, in which W_3 is on the front face of the structure, and the carboxylate anion binding pocket is indeed formed to an appreciable extent prior to binding the cell wall analogue. The interconvertion of the two conformers requires the rotation, through approximatety 180°, of the amide bond formed between the carbonyl of residue 2 and the NH of residue 3. This process is indicated schematically in 11.

The above process is slow on the NMR timescale, as deduced from the following evidence (33). At room temperature, the resonances of the amide NH protons W_2, W_3, and W_4, associated with residues 2, 3, and 4, are very broad compared with those (W_5, W_6, and W_7) of the other amide NHs (Fig. 16, spectrum obtained in d_6-DMSO solution). On heating the solution, the resonances of W_2 (initially at least), W_3, and W_4 sharpen, with those of W_3 and W_4 attaining approximately the same linewidth as those of W_5 and W_7 at @ 350°K (Fig. 16). Above 310°K, the exchange of W_2 with traces of water in the solvent becomes the primary determinant of the linewidth of its resonance; this process is incidental to our present considerations, but is catalysed by the basic NH_2 group at the N-terminus of the antibiotic. Similar, though less dramatic sharpening is observed for the resonances of X_2, X_3, and Z_2 but not for other α-CH and β-CH protons. These selective sharpenings of the specified lines indicate a process which is passing from an intermediate exchange rate to the fast exchange limit as the temperature is increased.

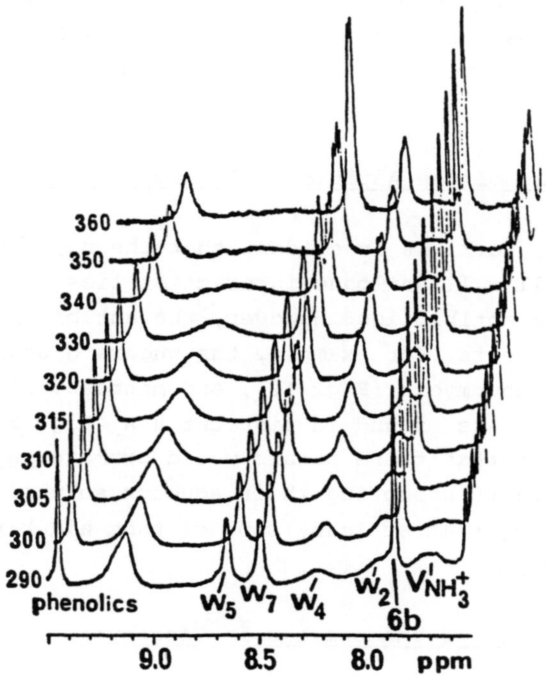

Fig. 16

In the light of the above nOe evidence, and the fact that the selectively sharpened resonances are all due to protons in the vicinity of the amide bond connecting residues 2 and 3, we conclude that this amide bond is undergoing slow rotation. Estimates of the chemical shift differences which might be observed in the slow exchange limit lead us to conclude that the amide bond is rotating at a frequency in the range 200-1000 Hz (33). Thus the barrier to the rotation must lie near to 12-14 kcal mol^{-1}.

The above conclusions may be of utility in the design of a carboxylate anion-binding pocket. They indicate that an appropriate pocket might be found within a tetrapeptide unit if the residues (counted from the N-terminus) had R, R, S, and R (or S, S, R, and S) configurations, and an appropriate connection of the sidechains of residues 2 and 4. Such an arrangement can apparently destabilise a normal peptide backbone geometry to such an extent that the free energies of three successive NHs 'up-down-up' and 'up-up-up' become comparable. It is moreover evident from our considerations that the desired 'up-up-up' arrangement can be further stabilised by appropriate cross-linking of residues 1 and 3 (at the cost of slightly greater synthetic complexity). The synthesis of such pockets is in hand.

(b) The Role of the Antibiotic Sugars in Binding

From the early work, it is evident that the major basis for the antibiotic-cell wall analogue interaction lies in the aglycone portion of the antibiotics. However, the antibiotics are found to carry 1 to 7 sugars. For example, the sugars glucose and vancosamine are found in vancomycin (Fig. 12), and mannose, ristosamine, and a tetrasaccharide are found in ristocetin A (Fig. 13). While it always appeared likely that these sugars would promote in vivo activity of the antibiotics by promoting their aqueous solubility, it is only now becoming evident that at least some of them aid binding in subtle and interesting ways.

A. The Role of Vancosamine in Vancomycin

When vancosamine is selectively removed from vancomycin, the binding constant to the tripeptide cell wall analogue drops by a factor of @ 3 (34). Inspection of an X-ray structure (Fig.11) of a degradation product of vancomycin, CDP-1, immediately suggests a reason for this drop. In the X-ray structure, the 6-methyl group of vancosamine lies immediately next to the benzene ring of the amino acid residue 4 (see (see Fig. 11); it is this benzene ring that undergoes a hydrophobic interaction with the methyl group of the C-terminal alanine residue of the cell wall analogue. Thus, the 6-methyl group of vancosamine can block the access of water to the methyl group of the C-terminal alanine residue, and so improve the overall hydrophobic interactions of the latter.

Further experiments show that although this idea is basically correct, the significance of the interaction is subtly modified by the presence of the NH_3^+ group of the amino-sugar vancosamine. Thus, acetylation of the amino group of vancosamine also reduces the binding constant to the same cell wall analogue by a factor of @ 3 (34). Determination of nOes when the cell wall analogue is bound to vancomycin, and when bound to vancomycin N-acetylated in the sugar residue, establishes that the 6-methyl group of vancosamine is adjacent to the methyl group of the C-terminal alanine residue in both complexes. Hence, we are forced to the conclusion that the associated hydrophobic interaction is strengthened (and indeed only contributes significantly to binding) in the presence of the charged amino group.

The physical basis for such an effect is clear. Water molecules in the vicinity of the NH_3^+ group will be relatively strongly bound to each other, and relatively highly ordered. As a consequence, the penalty in free energy to be paid for disrupting this local water structure by a hydrocarbon surface will be larger than in the absence of the charge. Thus, the hydrophobic effect is strengthened by the proximate charge. A macroscopic expression of this kind of effect is found in the familiar 'salting-out' phenomenon used in working up some organic reactions; a substance of limited hydrophobicity can be induced to pass from an aqueous to a non-polar phase by increasing the ionic strength in the aqueous phase. Of course, since the effect in the antibiotic is intramolecular, the local 'salting out' is highly effective.

B. The Role of Mannose in Ristocetin A

The removal of mannose from residue of ristocetin A (Fig.13) reduces the binding to N-acetyl-D-Ala-D-Ala by slightly in excess of 1kcal mol^{-1} (@ a factor of 10 reduction in the binding constant) (35). The molecular basis of this effect seems clear from earlier proton nOe experiments (36). The anomeric proton of the mannose residue shows an nOe to the methyl sidechain of the central alanine residue of the tripeptide cell wall analogue. Hence, it is concluded that a portion of this methyl group can be shielded from water by a relatively hydrophobic portion of the sugar residue.

However, the binding constant of ristocetin A to tripeptide is not changed significantly upon removal of mannose (and simultaneously of the tetrasaccharide) (37). It must be concluded that although the mannose residue is important in shielding the methyl group of the non-C-terminal alanine residue from water in dipeptide, it has no such significant role when the tripeptide binds. This apparent anomaly is resolved by the observation that when the tripeptide binds to ristocetin A in aqueous solution, the lysine sidechain folds over the aromatic ring of residue 7. Evidently, the lysine sidechain protects the methyl group of the non-C-terminal alanine from solvent to an extent that makes the corresponding role of mannose insignificant. It may additionally be that in the complex with tripeptide, the lysine sidechain and the mannose show an unfavourable steric interaction that prevents them from being able to shield the methyl group simultaneously and optimally from water.In this connection, it is noteworthy that vancomycin, which lacks the mannose attached to amino acid 5, binds much more strongly to tripeptide (1.5 x 10^6) than to dipeptide (6.4 x 10^4 mol^{-1}). Thus, mannose favors the binding of dipeptide, but not tripeptide. We conclude that mannose is present in ristocetin A to provide selectivity for binding to cell wall precursors having D-Ala at the penultimate position, i.e., to discriminate against sidechains other than methyl.

(iv) Study of the Peptide Secretion of the Clawed Toad Xenopus Laevis: the First Peptide Antibiotics from a Vertebrate

The studies which have been described so far have been in the traditional domain of natural products chemistry where those products are derived from bacteria, algae, fungi (probably) and plants. However, it seemed to us that there might also be a range of physiologically active natural products in animals, which so far have not been extensively examined. Following the advent of fast atom bombardment mass spectrometry in the early 1980s, we therefore began a search for peptides, which might have interesting physiological functions, in vertebrates. Prior to our work, it was well known that peptides which are produced in amphibian skin often have interesting physiological effects in mammals. Therefore, if a amphibian peptide of known structure produces a physiological effect in a mammal, and yet no mammalian peptide is known to give a similar physiological effect, it seems worthwhile to search for the mammalian analogue using the structure of the amphibian peptide to guide the search. Therefore, this strategy was first followed using the amphibian peptide kassinin. Antibodies were reared in the rabbit to this amphibian peptide, and these antibodies were then used to screen an extract of cow's spinal cord. The basis of this biological assay was that since the antibodies had been raised against a portion of the molecular structure of kassinin, these antibodies should be relatively specific in binding to any structurally analogous peptide from cow's spinal cord. The screen was successful, and led to the isolation by preparative HPLC of approximately 10^{-7} g of a peptide which appeared to contain methionine. However, the size of the sample was insufficient for a structure determination, and the peptide at this time was simply named substance K(38). Subsequent to this work, a determination of the c-DNA sequence which codes for the well-known neuropeptide substance P showed that the c-DNA of the precursor of substance P would also code for a peptide having the same C-terminal sequence as kassinin(39). This was clearly therefore substance K, and subsequent synthesis of the appropriate sequence has shown that this is indeed a new neuropeptide which does have receptors in brain which are distinct from those found for substance P.

Encouraged by the success of the above strategy, we decided that it was important to characterise as fully as possible peptides found in the skin secretion of the frog Xenopus laevis, not least because any newly determined structures could be used in the same way as above to search for new neuropeptides in mammals. Secretion of peptides through the skin of the frog is induced by injection of adrenalin into a dorsal sac situated near the middle of the frog's back. A mass spectrometric study of the secreted peptides led to the identification of no less than 30 new peptides in the skin secretion(40). Two strategies (40,41) were used to determine the sequences of these peptides, and these are now outlined.

The first strategy(41) is applied to peptides whose sequence must be determined in toto, since in these cases no DNA sequence information is available to complememnt the mass spectrometric sequencing work. This approach utilises enzymic digestions to provide sequence information which, due to incomplete fragmentation in FAB MS, would otherwise be missing; it is illustrated by the sequence determination of the frog peptide laevitide. Purification of the skin secretion of the frog by preparative HPLC gave a peptide, laevitide, whose molecular weight was determined to be 1541 from the positive ion fab mass spectrum. Fragment ions gave the sequence of the first four amino acids from one end of the peptide, with the qualification that the first residue was the amide of lysine or glutamine, the second residue was lysine or glutamine, and the third residue leucine or isoleucine (since the masses of these pairs of amino acids are the same). When the molecular weight of the peptide was considered in conjunction with its amino acid analysis, it became clear that the constituent amino acids gave a mass which was 18 mass units greater than that observed in the natural product. This was indicative of the presence of pyroglutamic acid at the N-terminus, and this in turn implied that the sequence obtained as above corresponded to the sequence at the C-terminus. Given that arginine is thereby determined to be the fourth residue in from the C-terminus of the peptide, it was possible to cleave off the C-terminal 3 peptides with the enzyme trypsin (which cleaves at the C-terminal side of arginine and lysine residues). When this enzymic digestion of the peptide was carried out, it was found that the FAB mass spectrum of the product not only contained the peptide of expected molecular weight 1174, but also a similar amount of a peptide of molecular weight 1018. The mass difference of these two peptides corresponds to the mass of arginine, and this indicates that the arginine may also be removed in the digestion. Thus it must be concluded that the residue on the N-terminal side of arginine is lysine (since amino acid analysis shows the presence of only one arginine residue). Sequence ions were obtained from the peptide of molecular weight 1018 and they established the sequence PyroGluGlyMetLeu (or Ile) at the N-terminus.

More information as to the structure of the peptide of molecular weight 1018 was obtained by digesting this material with the enzyme carboxypeptidase Y. This enzyme removes residues from the C-terminus of peptides, and indeed it was found that the product from this digestion contained peptides of molecular weight 890, 803, and 702. The mass differences of these peptides from that of the parent molecular weight of 1018 indicates the successive removal of lysine (mass 128), serine (mass 87), and threonine (mass 101). The sequence determined so far is that summarised in structure 12.

PyroGlu-Gly-Met-Leu ..... Thr-Ser-Lys-Arg-Leu-Lys-Gln-NH_2

(Ile) (Ile)(Gln)(Lys)

12

To determine the remaining part of the sequence, it was necessary to locate the position of one leucine in the molecule, which had to be differentiated from two isoleucines known to be present from the amino acid analysis. Since the enzyme chymotrypsin is selective for cleavage at the C-terminal side of leucine, a digestion with this enzyme should in principle allow the determination to be made. In fact, the product of the chymotryptic digestion contained a peptide of molecular weight 702, which corresponds to the loss of the C-terminal 7 amino acids from laevitide. This product of molecular weight 702, could in turn then be subject carboxypeptidase Y digestion, when two new products of molecular weight 589 and 488 were obtained. These two fragments correspond to the loss of leucine and threonine, respectively, from the parent molecule of molecular weight 702. When this new information is added to that already summarised in 12, it became clear that the only residue missing was that of glycine (both from the amino acid composition and from the undetermined mass). Additionally, it can be shown by FAB MS that the penultimate residue (but not the C-terminal residue) can be acetylated by acetic anhydride, and therefore lysine is the penultimate residue. Thus, the full structure is determined as that shown in 13. This structure was confirmed by solid phase peptide synthesis, which gave a product which co-chromatographed with the natural material.

PyroGluGlyMetIleGlyThrLeuThrSerLysArgIleLysGln-NH_2

13

The second strategy(40) is applied to those peptides for which the DNA sequence of the precursor is known. In this case, the determination of a partial peptide sequence by FAB MS allows this partial sequence to be matched against the sequence of the prepropeptide predicted from the DNA sequence. In the next step, the molecular weight of the peptide is used to establish the size of the peptide which is cleaved from the prepropeptide. For example, a portion of the preprocaerulin cDNA sequence determines the amino acid sequence shown in Fig. 17.

Portion of Preprocaerulin cDNA and amino acid sequence

-----CTC CGA GGA TTT GGC TCT TTC CTA GGT AAA GCT

Leu Arg Gly Phe Gly Ser Phe Leu Gly Lys Ala

112 →

TTA AAG GCT ----- MH^+ 996

Leu Lys Ala

123

Fig. 17

The partial sequence Gly-Phe-Gly-Ser-Phe- of a secreted peptide, and its molecular weight of 995, determined by FAB MS are sufficient to establish the complete sequence:

Gly-Phe-Gly-Ser-Phe-Leu-Gly-Lys-Ala-Leu (see Fig. 17)

Among the peptides whose sequences were determined in the above ways were many that had the ability to lyse cell membranes; either membranes which had been reconstituted artificially from known membrane components, or the natural membranes of red blood cells. Particularly interesting among these were two peptides, each containing 23 amino acids, which have the sequences shown in 14 and 15. These peptides were named PGS and [Gly-10,Lys-22]PGS.

```
Gly-Ile-Gly-Lys-Phe-Leu-His-Ser-Ala-Lys-Lys-Phe
1                                   10

Gly-Lys-Ala-Phe-Val-Gly-Glu-Ile-Met-Asn-Ser
                                    22
```

14

```
Gly-Ile-Gly-Lys-Phe-Leu-His-Ser-Ala-Gly-Lys-Phe
1                                   10

Gly-Lys-Ala-Phe-Val-Gly-Glu-Ile-Met-Lys-Ser
                                    22
```

15

We noted that the peptides 14 and 15, if coiled into α-helices (of the kind that occur as portions of the structures of proteins), give rise to cylindrical structures in which one side of the helix contains many polar side chains and the other side of the helix contains many hydrocarbon side chains. Such helices are called amphiphilic. It appeared likely that this amphiphilic structure is relevant to the membrane disrupting properties of the peptide, since the length of the cylinder is just sufficient to span a membrane. For example, it is possible that the aggregation of several such cylinders within the membrane would give rise to a pore out of which key contents of the cell could escape (see Fig. 18) thus providing a mechanism for cell death.

Our discovery and sequence determination of these peptides was made much more exciting by their independent sequencing by Zasloff(42), who also described that the peptides have antibacterial activity. Whether these peptides will eventually find a clinical role as antibacterial substances is still open to question, but it is clear that the two independent pieces of work have described a peptide defence system, independent of the immune system, in a vertebrate.

The precise nature of the above defence system is still open to debate. In our own work, we had noted that the secretion of lytic peptides through the back of the frog, and also of hormonally active peptides (such as gastrin) which are also contained in the secretion, would provide the frog with a defence system against predators. Clearly, large amounts of a gastric hormone, plus lytic peptides, taken internally by a predator would cause it to experience unpleasant effects. In addition, Zasloff has reported evidence that PGS and [Gly-10,Lys-22]PGS (which he calls respectively magainin 2 and magainin 1) are present in subdermal fluids in the frog, and thus may be part of an antibacterial defence system. Indeed, he was guided in his approach by the fact that these frogs when treated surgically to remove their eggs (for other biological studies) could, after suturing of the wound, be returned to non-sterile water without subsequent infection of the wound. How such peptides may circulate internally in the frog without disrupting its own membranes, yet have the ability to disrupt membranes in bacteria, fungi, and protozoa is a question of enormous interest and of great importance with regard to any clinical potential these peptides might have.

Conclusion

It will be clear from the foregoing sections that an extremely wide variety of organisms produce peptides which appear to be used in the battle for survival. It is also clear that we now have powerful tools to elucidate the structures of these substances and, in the cases where their receptors can be isolated as low molecular weight entities, that nmr can be used to determine the molecular basis for their physiological effects. Although those which are produced as toxins appear to have little in the way of clinical potential, it is

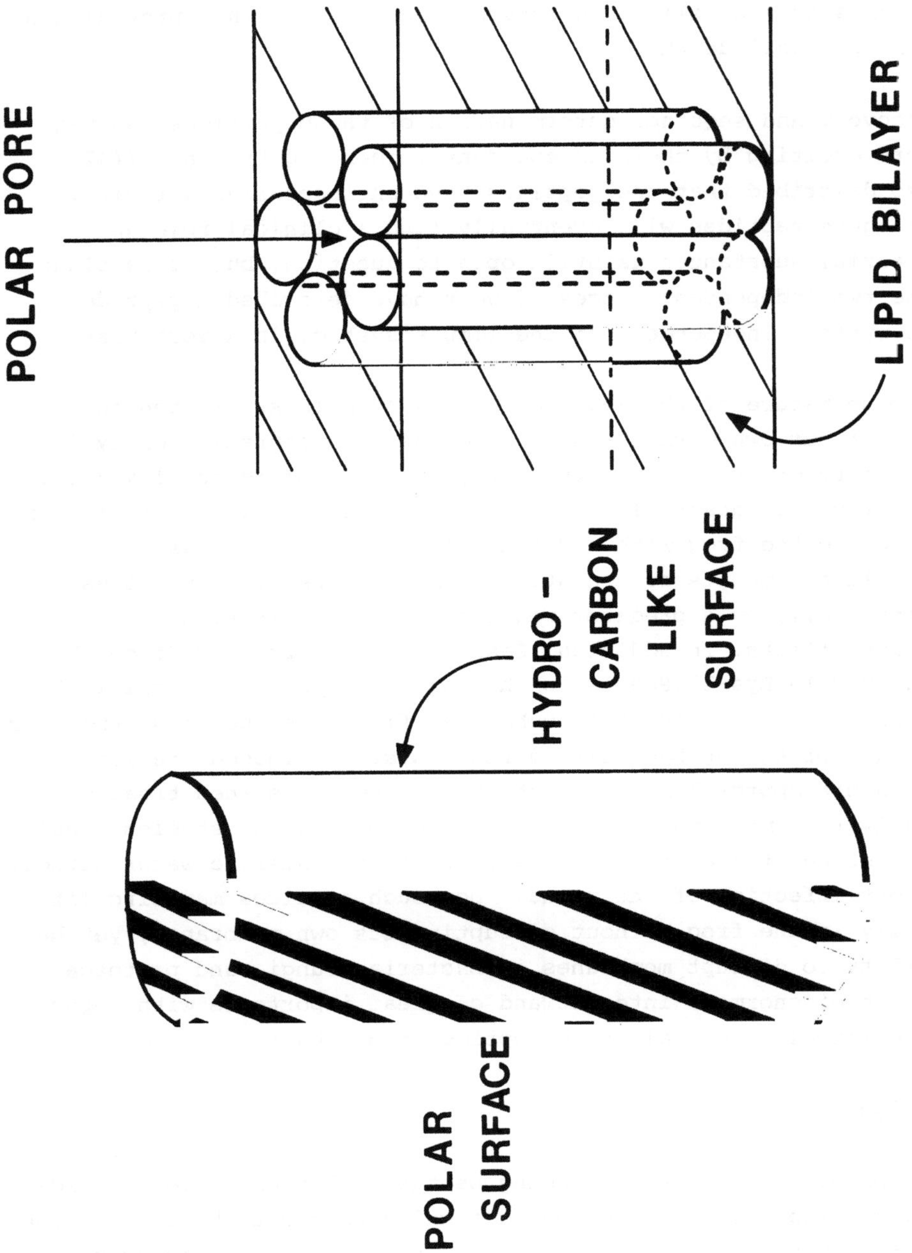
POLAR PORE
LIPID BILAYER
HYDRO –
CARBON
LIKE
SURFACE
POLAR
SURFACE

Fig. 18

clear that an ongoing search for peptides in both actinomycetes and in higher animals is probably a well worth exercise. It may well be that many of the important drugs of tomorrow will be found from such sources.

References

1 Budzikiewicz H, Djerassi C, Williams DH (1967) Mass Spectrometry of Organic Compounds, Holden-Day, San Francisco

2 Barber M, Bordoli RS, Sedgwick RG, Tyler AN (1981) J Chem Soc Chem Commun 325

3 Williams DH, Santikarn S, De Angelis F, Smith RJ, Reid DG (1983) J Chem Soc Perkin Trans I 1869

4 Radhakrishna AS, Parham ME, Riggs RM, Loudon GM (1979) J Org Chem 44:1746

5 Gil-Av E, Feibush B (1967) Tetrahedron Letters 3345

6 Dabrowiak JC, Cooke DW (1971) Anal Chem 43:791

7 Liardon R, Ledermann S, Ott U (1981) J Chromatogr 203:385

8 Kusumoto S, Matsukura M, Shiba T (1981) Chem Lett 1117

9 Iwase H, Murai A (1974) Chem Pharm Bull 22:8

10 Holcomb GN, James SA, Ward DN (1968) Biochemistry 7:1291

11 Botes DP, Tuinman AA, Wessels P, Viljoen CC, Kruger H, Williams DH, Santikarn S, Smith RJ, Hammond SJ (1984) J Chem Soc Perkin I 2311

12 Leung TWC, Williams DH, Barna JCJ, Foti S (1986) Tetrahedron 42:3333

13 Doddrell DM, Pegg DT, Bendall MR (1982) J Mag Res 48:323

14 McCormick MH, Stark WM, Pittinger GF, Pittinger RC, Mcguire GM (1955-56) Antibiotics Ann 606

15 Tedesco F, Markham R, Gurwith M, Christie D, Bartlett JG (1978) Lancet 226

16 Marshall FJ (1965) J Med Chem 8:18

17 Johnson CR (1969) Thesis, University of Illinois

18 Williams DH, Kalman JR (1977) J Am Chem Soc 99:2768

19 Sheldrick GM, Jones PG, Kennard O, Williams DH, Smith GA (1978) Nature 271:223

20 Williamson MP, Williams DH (1981) J Am Chem Soc 103:6580

21 Harris CM, Kopecka H, Harris TM (1983) J Am Chem Soc 105:6915

22 Williams DH, Rajananda V, Kalman J (1979) J Chem Soc Perkin I 1695

23 Kalman J, Williams DH (1980) J Am Chem Soc 102:897

24 Fesik SW, Armitage IM, Ellestad GA, McGahren WJ (1982) 184th Natl. Meet. Am. Chem. Soc. Kansas City Abstr MEDI 66 Washington DC: Am Chem Soc
25 Hunt AH (1983) J Am Chem Soc 105:4463
26 Debono M, Merkel KE, Weeks R, Cole HJ (1982) US Patent 4,322,406; and references therein
27 Barna JCJ, Williams DH, Stone DJM, Leung TW, Doddrell DM (1984) J Am Chem Soc 106:4895
28 Hammond SJ, Williamson MP, Williams DH, Boeck LD, Marconi GG (1982) J Chem Soc Chem Commun 344
29 Hammond SJ, Williams DH, Nielsen RV (1983) J Chem Soc Chem Commun 116
30 Perkins HR (1969) Biochem J 111:195
31 Brown JP, Feeney J, Burgen ASV (1975) Mol Pharmacol 11:119
32 Kalman J, Williams DH (1980) J Am Chem Soc 102:906
33 Williams DH, Waltho JP (1988) Biochem Pharm 37:133
34 Kannan R, Harris CM, Harris TM, Waltho JP, Skelton NJ, Williams DH (1988) J Am Chem Soc 110:2946
35 Barna JCJ, Williams DH, Williamson M (1985) J Chem Soc Chem Commun 245
36 Williams DH, Williamson MP, Butcher DW, Hammond SJ (1983) J Am Chem Soc 105:1332
37 Herrin TR, Thomas AM, Perun TJ, Mao JC, Fesik SW (1985) J Med Chem 28:1371
38 Maggio JE, Sandberg BEB, Bradley CV, Iverson LL, Santikarn S, Williams DH, Hunter JC, Stanley MR (1983) Irish J Med Sci 152 (suppl 1):45
39 Nawa H, Kotani H, Nakanishi S (1984) Nature 312:729
40 Gibson BW, Poulter L, Williams DH, Maggio JE (1986) J Biol Chem 261:5341
41 Poulter L, Terry AS, Williams DH, Giovannini MG, Moore CH, Gibson BW (1988) J Biol Chem 263:3279
42 Zasloff, M (1987) Proc Natl Acad Sci USA 84:5443

Genetically Designed Affinity Chromatography Using a Novel Metal Chelate Absorbent

E. Hochuli

Hoffmann-La Roche & Co. ZFE/MB
CH-4002 Basel

U. P. Schlunegger (Ed.)
Biologically Active Molecules

INTRODUCTION

Modern molecular biology has made it possible to produce many biologically interesting proteins in bacteria, yeast or mammalian cells. However, when recombinant proteins are expressed in host cells and subsequently grown in fermentation tanks, the desired product must be recovered in pure form. Protein purification by traditional multi-step chromatogrsphic separation is tedious and affords low overall yields.

Recently very attractive approaches to facilitate purification of recombinant proteins have been developed. Hybrid proteins were prepared by fusing the coding sequence of the protein of interest with the coding sequence of a peptide with high affinity for an affinity resin together with the sequence of a specific cleavage site. Such fusion proteins could efficiently be purified, taking advantage of the specific binding of the affinity peptide (affinity tag) to the affinity resin. After purification of the fusion protein, the affinity tag could be split off at the designed cleavage site. One example described in literature is the use of fusions with protein A, e.g. human insulin-like growth factor, where separation is achieved through the interaction of protein A with an IgG affinity column (1). Other gene fusion systems, based on fusion to ß-galactosidase (2) and polyarginine (3) have been reported. The ß-galactosidase hybrid proteins have been purified on a ß-galactosidase-specific affinity resin and the polyarginine hybrid protein on a cation exchange resin.

In this article we shall describe a new genetically designed protein purifiaction system. It is based upon the selective interaction between a poly-histidine peptide, which is fused to the protein of interest, and a novel metal chelate adsorbent. The principle of the technique is illustrated with mouse dihydrofolate reductase. DNA elements coding for adjacent histidines were fused to the mouse dihydrofolate reductase gene. Subsequent expression in E.coli resulted in the production of hybrid proteins which could be purified by immobolized metal ion affinity chromatography. Furthermore the removel of the histidine affinity peptide with carboxypeptidase A is illustrated.

CONCEPT OF PROTEIN PURIFICATION USING A GENETIC APPROACH

The use of genetic engineering techniques has allowed the synthesis of hybrid proteins. By fusing the coding sequence of a protein of interest with the coding sequence of a polypeptide with high affinity to a ligand, a hybrid protein with an affinity tag can be produced directly by a microorgansism. The affinity tag can then be used to recover the product from a culture medium or a cell lysate by affinity chromatography. It is also possible to introduce a specific cleavage site at the junction between the protein parts which enables cleavage of the hybrid molecule to yield the protein of interest free of affinity tag.

The concept is schematically shown in Fig. 1. A crude extract containing a hybrid protein consisting of the desired potein X fused with an affinity tag A is passed through an affinity column containing a specific ligand L (1). Only the protein molecules which have an appreciable affinity for the immobilized ligand will be retained (2) and all the contaminants are washed out (3). After elution, the purified fusion protein can be cleaved at the designed cleavage site (4) to give the desired protein X and the affinity tag A which can then be removed by a second passage on the affinity column. This purification protocol offers advantages over traditional recovery processes and it might provide a general procedure for the purification of recombinant proteins.

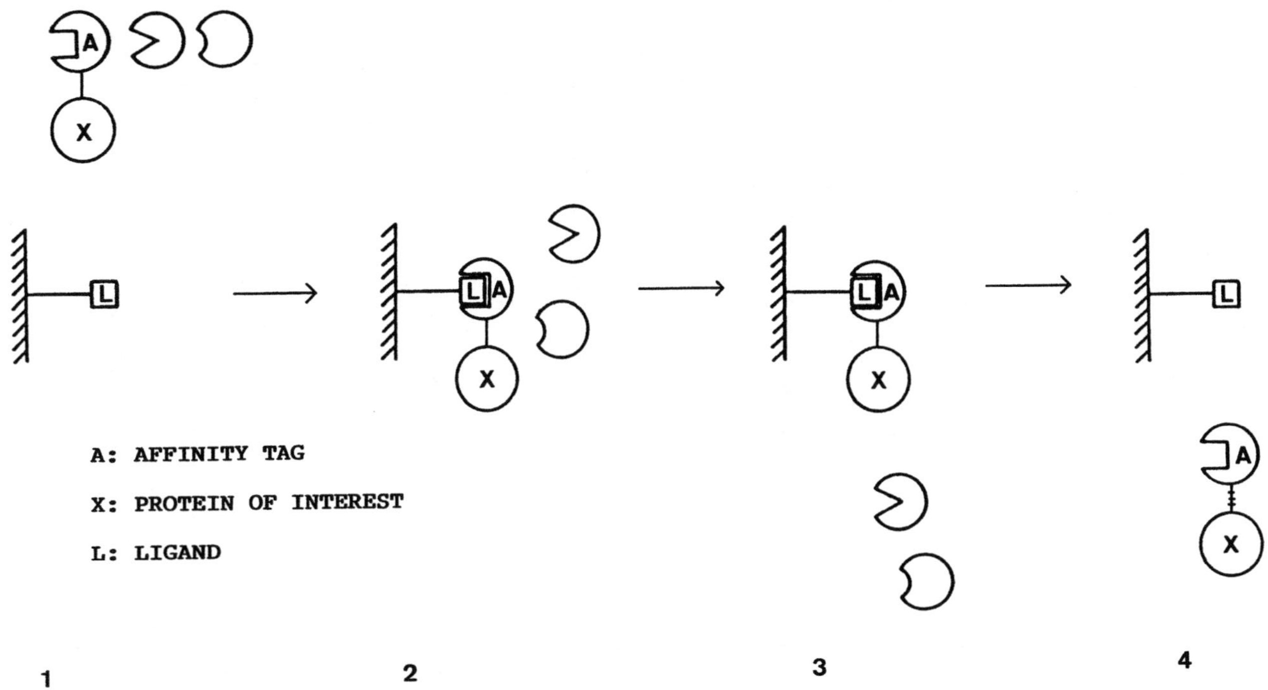

Fig. 1: Schematic representation of the protein purification system using a genetic approach.

Cleavage Of Fusion Proteins

To obtain sequence specific cleavage of a hybrid protein it is possible to use either a chemical or an enzymatic method. Fig. 2 shows some possibilities. The utility of chemical cleavage of proteins is limited since only short stretches of amino acids are recognized. If the desired protein contains the recognized amino acid or dipeptide sequence, it will be degraded during cleavage. In contrast, enzymatic digestion is more attractive for general applications since proteases often cut at the C-terminal of a specific recognition site. However, enzymatic cleavage is highly dependent on steric factors which might prevent the protease from cleaving the recognition site. A further disadvantage is possible contamination by other proteases which might nonspecifically degrade the fusion protein resulting in difficulty in purifying the desired protein. In this respect, chemical agents are advantageous since they can be easily separated from the cleavage mixture by dialysis or a simple desalting step.

Chemical:	
Cyanogen bromide	Met ↓
Formic acid	Asp ↓ Pro
Hydroxyl amine	Asn ↓ Gly
Oxidative reagents	Trp ↓
Enzymatic:	
Enterokinase	Asp Asp Asp Lys ↓
Factor Xa	Ile Glu Gly Arg ↓
Collagenase	... Pro Val ↓ Gly Pro ...

Fig. 2: Cleavage sites to cleave gene fusion products

From the preceeding discussion it is evident that the method of cleavage must be determined on a case-by-case basis, after careful analysis of the amino acid sequence of the protein of interest.

IMMOBILIZED METAL CHELATE AFFINITY CHROMATOGRAPHY

Immobilized metal chelate affinity chromatography is a well known technique for protein purification. Porath et al. showed that immobilized cations are very useful for the chromatographic purification of proteins (4). The chelating ligands iminodiacetic acid (IDA) and tris-(carboxymethyl)ethylene diamine (TED) have been bound covalently to oxirane-activated agarose and the resulting gels charged with metal ions as Cu^{2+}, Zn^{2+} or Ni^{2+} (5). These adsorbents have since been used successfully for the purification of many peptides and proteins (6).

When chelated in the tridentate IDA resin, Cu^{2+} ions leave one coordination site free for interactions with biopolymers, whereas Ni^{2+} ions leave three (fig. 3a). However, Ni^{2+} ions are not bound tightly enough and releasing of metal ions from the adsorbent may be observed during chromatography.

The pentadentate TED resin binds metal ions with coordination number six very strongly, but again only one coordination site remains for biopolymer interaction (fig. 3b).

An ideal adsorbent for immobilized metal affinity chromatography must on the one hand strongly complex the metal ions and on the other hand must permit reversible interactions between immobilized metal ions and proteins. Immobilized iminodiacetic acid (IDA resin) largely fulfils these requirements for Cu^{2+} ions but only to a limited extent for Ni^{2+} or Zn^{2+} ions, since the latter are only weakly bonded. Ni^{2+} ions are washed out from IDA resin upon loading with strong chelating proteins or peptides as e.g.

His-His-Leu-Gly-Gly-Ala-Lys-Glu-Ala-Gly-Asp-Val.
On the other hand, Ni^{2+} or Zn^{2+} chelate adsorbents are of particular interest, as Ni^{2+} and Zn^{2+} have the power to complex six ligands, whereas Cu^{2+} ions preferably bind four in the inner coordination sphere.

a

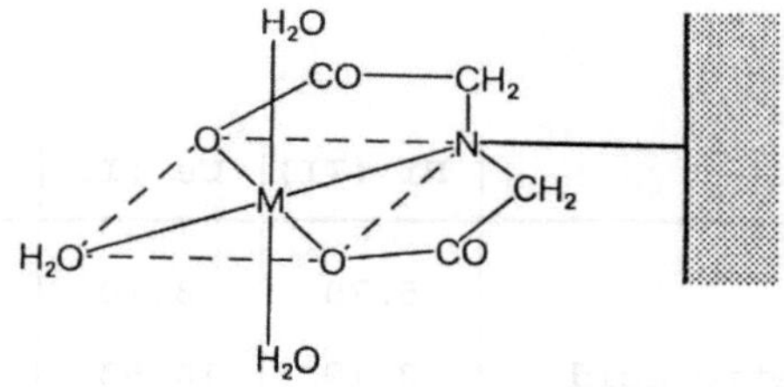

b

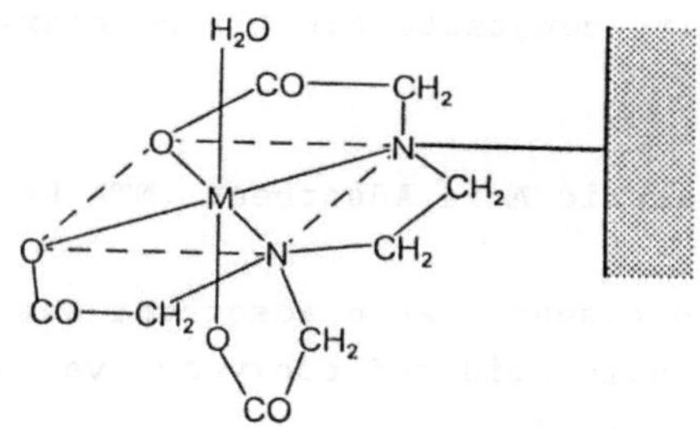

Fig. 3: Schematic structure of metal chelate adsorbents charged with metal ions with coordination number 6 (Ni^{2+})

a) The tridentate iminodiacetic acid adsorbent (IDA resin): Three valencies are occupied with H_2O and remain for biopolymer interaction
b) The pentadentate tris(carboxymethyl) ethylene diamine adsorbent (TED resin): One binding site is occupied with H_2O and remains for biopolymer interaction.

The stability constants for 1:1 complexes of different soluble metal-chelates (7) (see fig. 4) demonstrate that the nitrilotriacetic acid complexes are much more stable than the imminodiacetic acid complexes. From these stability constants it can be concluded that a nitrilotriacetic acid adsorbent (NTA resin) for immobilized metal chelate affinity chromatography would be more stable against metal ion leaching than the known IDA resin.

Ligand	Ni (II)	Cu (II)	Zn (II)
Glycine	5.70	8.10	5.20
Iminodiacetic acid	8.19	10.63	7.27
Nitrilotriacetic acid	11.50	13.00	10.70
Imidazole	3.01	4.26	2.13

Fig. 4: Stability constants for 1:1 complexes (log K)

The Nitrilotriacetic Acid Adsorbent (NTA Resin)

The new nitrilotriacetic acid adsorbent was synthesized with lysine, bromoacetic acid and epoxy-activated Sepharose Cl-6B as outlined in fig. 5 (8).
The ligand content in the chelating resin and the amount of metal ions, Cu^{2+} and Ni^{2+}, retained by the resin after washing with 0.2M acetic acid, are given in fig. 6.
From the excellent agreement of the ligand concentration with the metal ion concentration it can be concluded that in the adsorbent a 1:1 metal ligand complex is formed.

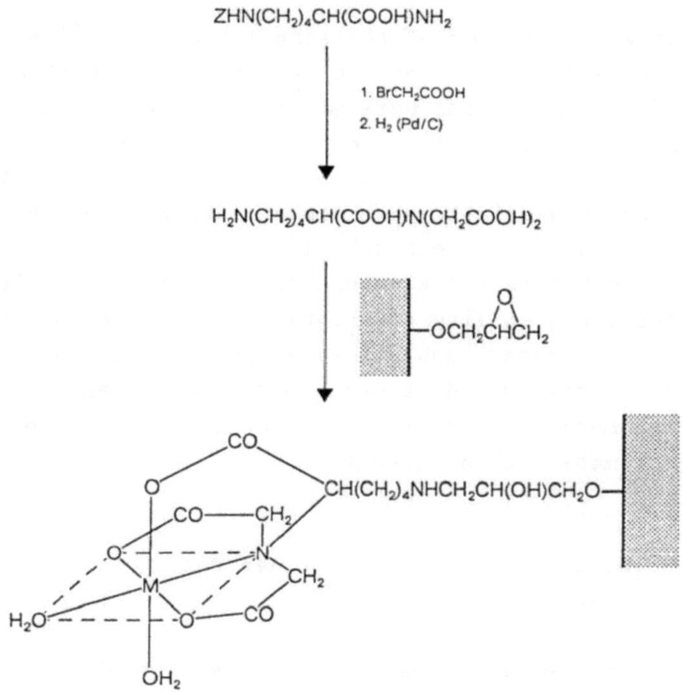

Fig. 5: Preparation of the new quadridentate nitrilotriacetic adsorbent (NTA resin) charged with metal ions with coordination number 6

	μequivalent/ml adsorbent
Ligand	9.5
Cu^{2+}	9.4
Ni^{2+}	7.1

Fig. 6: Ligand content, determined by nitrogen elemental analysis and metal content, determined by atomic adsorption spectroscopy of the new NTA adsorbent

For a qualitative comparison of the stabilities of Cu^{2+} and Ni^{2+} NTA and IDA adsorbents, the two resins charged with a mixture of Cu^{2+} and Ni^{2+} ions were eluted with an aqueous solution of 0,1M iminodiacetic acid pH 7 and the washing out of the metal ions was followed. It can be assumed that the metal ions were eluted sequentially according to their strength of sorption on the resin and their relative affinity for iminodiacetic acid. Therefore the elution volume of the metal ion is a criterion for the stability of the metal chelate adsorbent. As shown in fig. 7 the metal ions are retained stronger on the NTA adsorbent than on the IDA adsorbent. Therefore it can be concluded that the metal NTA adsorbent is, as expected to be, more stable than the metal IDA adsorbent.

Affinity Of Selected Proteins And Peptides For The Ni-NTA Adsorbent

The affinity of proteins and peptides for metal chelate adsorbents is mainly due to interaction of surface exposed histidine residues with the immobilized metal ions (9).
In fig. 8 the affinity of selected histidine residues containing proteins for the $Ni^{2\pm}$ NTA adsorbent are shown.
Cytochrome C from horse heart, trypsin from bovine pancreas and interferon α-2a synthesized in bacteria were used as comparative substances. None of these proteins bonded to the adsorbent at pH 8. Obviously the arrangement of the histidines plays a decisive role. In the case of trypsin and interferon α-2a, three histidine residues are situated in position 29, 46, 79 (trypsin) and 7, 34, 57 (interferon) respectively. Obviously these histidines are not in the position to form a stable complex. In the case of cytochrome C the two histidines are spatially neighbouring, position 18 and 26, but cannot form a two-pronged ligand, since one histidine is bonded to the haem iron.

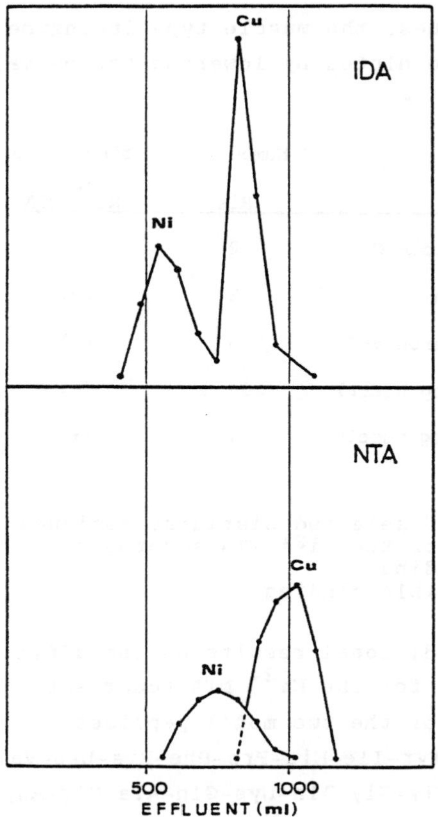

<u>Fig. 7:</u> Qualitative stability of the metal NTA adsorbent compared to the metal IDA adsorbent: A mixture of Cu^{2+} and Ni^{2+} ions was loaded on an IDA column and on a NTA column of identical size (160 ml) and metal binding capacity (10 µequivalent/ml). The columns were eluted with 0.1 M iminodiacetic acid pH 7. The metal ions are retained stronger on the NTA adsorbent than on the IDA adsorbent.

Lactate dehydrogenase isoenzymes are tetrameric proteins with a molecular weight of 140'000. The isoenzymes from hogs are largely homologuous with the exception of the amino terminal region. This is situated on the protein surface. The heart-type isoenzyme has no histidine in this region, but the muscle-type has three, among them the sequence -<u>His</u>-Asn-Leu-Leu-Lys-Glu-Glu-<u>His</u>-Val-Pro-<u>His</u>- (10).

While the heart-type enzyme was not adsorbed in spite of its

28 histidine residues, the muscle-type isoenzyme was adsorbed at pH 7,5 and could be eluted by lowering the pH values to 6.

Protein	Number of His	binding to Ni^{2+} NTA
Cytochrome C	2	(-)
Trypsin	3	(-)
Interferon α-2a	3	(-)
LDH (hog heart)	28	(-)
LDH (hog muscle)	31	(+)

Fig. 8: Affinity of selected histidine residues containing proteins for the Ni^{2+}-NTA adsorbent
(-) no binding
(+) reversible binding

In order to get additional results on the affinity of neighbouring histidines for the Ni^{2+} NTA adsorbent, the chromatographic behaviour of the two model peptides

A: Asp-Arg-Val-Tyr-Ile-His-Pro-Phe-His-Leu-Val-Ile-His-Ser

B: His-His-Leu-Gly-Gly-Ala-Lys-Gln-Ala-Gly-Asp-Val

was studied.

In a first experiment the peptides were chromatographed separately. 2.7 mg of each were loaded at pH 7 on a 29 ml column and eluted with a pH gradient from 5.8 to 4.8. Peptide A was eluted at pH 5.4 whereas peptide B was recovered at pH 5.25.

Fig. 9 shows the separation of a 1:1 mixture of the two peptides on the Ni^{2+}- NTA adsorbent under the identical chromatographic conditions. B having only two histidine residues in position 1 and 2 binds stronger than A having three histidine residues in position 6, 9 and 13.

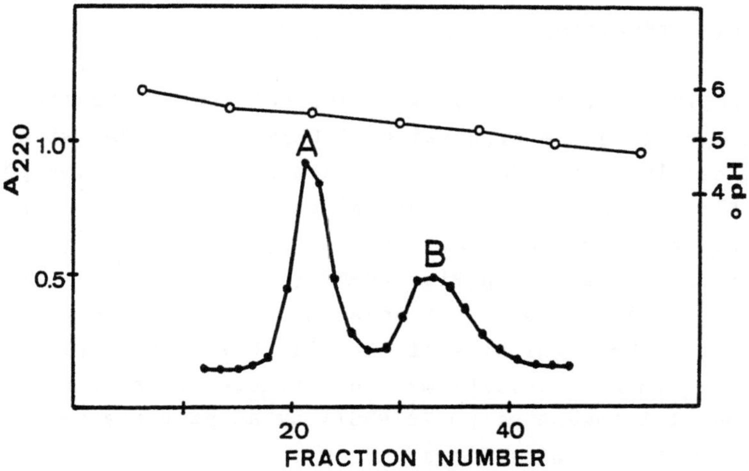

A Asp-Arg-Val-Tyr-Ile-His-Pro-Phe-His-Leu-Val-Ile-His-Ser

B His-His-Leu-Gly-Gly-Ala-Lys-Gln-Ala-Gly-Asp-Val

Fig. 9: Chromatography of the model peptides A and B on the Ni^{2+} NTA adsorbent. Elution was done with a gradient from pH 5.8 to 4.8.

These experiments indicate that the new Ni^{2+} NTA-adsorbent is selective for proteins and peptides which have as a structural element neighbouring histidines on the surface. Based on this remarkable specificity of the Ni^{2+}-NTA adsorbent a new genetically designed affinity chromatography system was developed (13).

The fusion concept is exemplified by mouse dihydrofolate reductase (DHFR). We prepared DNA elements which upon integration into expression vectors directed the synthesis of DHFR derivatives comprising two to six histidine residues, either at the amino terminus or at the carboxy terminus. To investigate the potential of poly-histidine fusions for protein purification, their chromatographic behaviour was studied under a variety of conditions. In addition, the removal of the histidine affinity-peptide after purification of the fusion protein with carboxypeptidase A (CPA) is demonstrated.

CLONING AND EXPRESSION OF THE FUSION PROTEINS

Plasmid pDS78/RBSII (fig. 10) was used for the construction of plasmids directing the expression of DHFR derivatives with adjacent histidine residues. This plasmid belongs to the pDS-family of plasmids (11) and contains the coding region for mouse dihydrofolate reductase (DHFR) from cell line AT-3000 (12). For construction of plasmids p(His)$_n$-DHFR (n = 2,3,4,5,6) adaptor fragments, encoding two to six adjacent histidines, were inserted into the BamHI site of plasmid pDS78/RBSII. Plasmids pDHFR-(His)$_n$ (n = 2,3,4,5,6) were constructed by replacing the BglII/HindIII fragment of pDS78/RBSII by adaptor fragments encoding two to six histidines.
E.coli cells were transformed with the different plasmids represented in fig. 10 and subsequently grown in medium containing ampicillin to select for the presence of the expression plasmid. After induction with IPTG the cellular proteins of the different cultures were analyzed by SDS-Page.

PURIFICATION OF THE FUSION PROTEINS

DHFR with two histidine residues at the carboxy terminus (DHFR-(His)$_2$) produced with cultures of E.coli was extracted by sonication in 0.05M phosphate buffer pH 8 and purified by chromatography on the Ni^{2+}-NTA column (fig. 11). SDS-Page and enzyme activity analysis demonstrate that DHFR-(His)$_2$ was retained at pH 8 on the adsorbent and the contaminating E.coli proteins were washed out. The fusion protein was eluted with a linear pH gradient from 8 to 5, yielding a product with a purity of at least 90% and a specific activity of 9 units/mg.
When DHFR containing six histidine residues at the amino terminus ((His)$_6$-DHFR) was extracted and purified in the phosphate buffer system, the fusion protein was retained on the

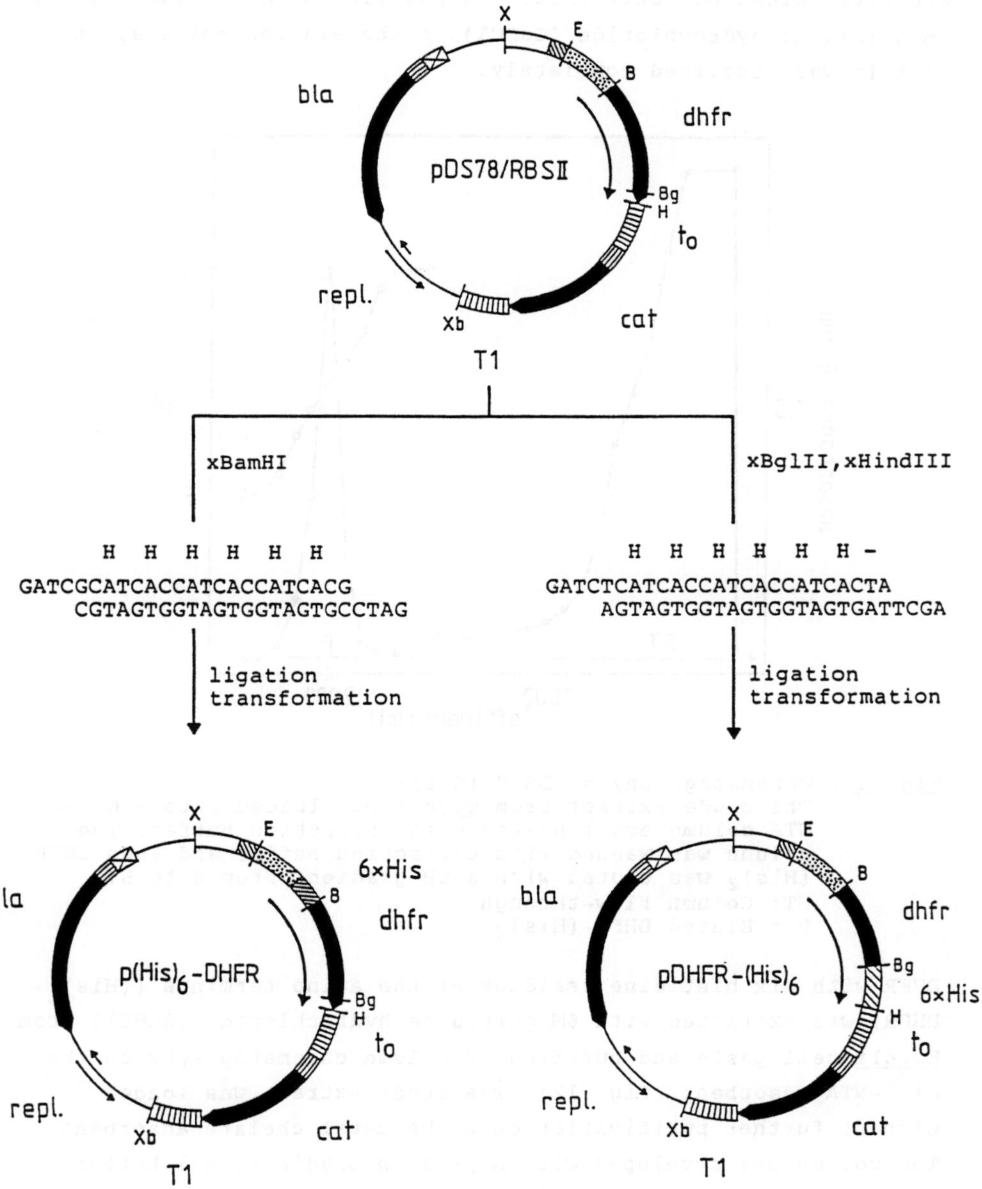

<u>Fig. 10:</u> Construction of plasmids directing the expression of DHFR derivatives with adjacent histidine residues.

affinity column but only about 10% was eluted. After addition of 6M guanidine hydrochloride (GuHCl) to the elution buffers, the protein was recovered completely.

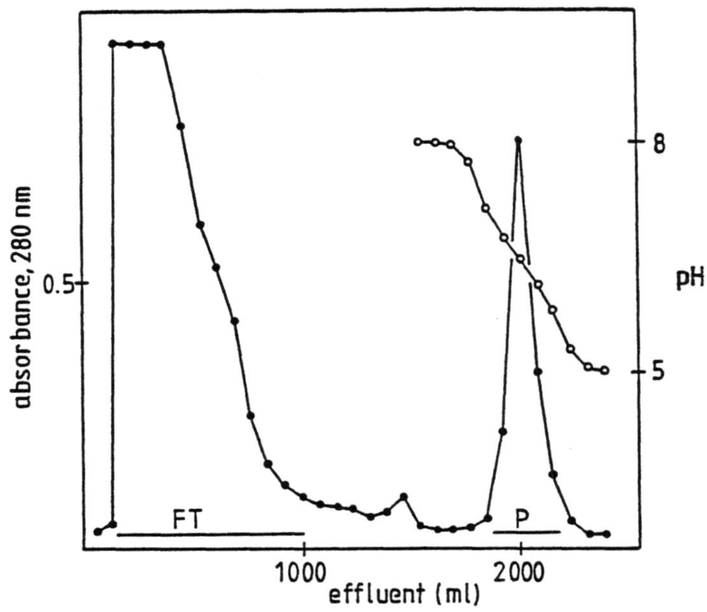

Fig. 11: Chromatography of DHFR-$(His)_2$
The crude extract from E.coli was loaded onto a Ni^{2+}-NTA column equilibrated with extraction buffer. The column was washed with extraction buffer and then DHFR-$(His)_2$ was eluted with a pH gradient from 8 to 5.
FT: Column flow-through
P : Eluted DHFR-$(His)_2$

DHFR with six histidine residues at the amino terminus ($(His)_6$-DHFR) was extracted with 6M guanidine hydrochloride (GuHCl) from E.coli cell paste and purified by column chromatography on the Ni^{2+}-NTA adsorbent (fig. 12). The crude extract was loaded without further purification onto the metal chelate adsorbent. The column was developed with a pH-step gradient. All buffer solutions contained 6M GuHCl throughout the chromatography. Collected fractions were dialyzed against water, lyophilized and analysed. SDS-Page analysis of the crude extract and the purified protein shows that the fusion protein was retained on the column at pH 8 and, after washing out the contaminants, was eluted with a purity of at least 90%.

Starting with 2 g of biomass, 25 mg $(His)_6$-DHFR with a specific activity of 2.3 units/mg were obtained.

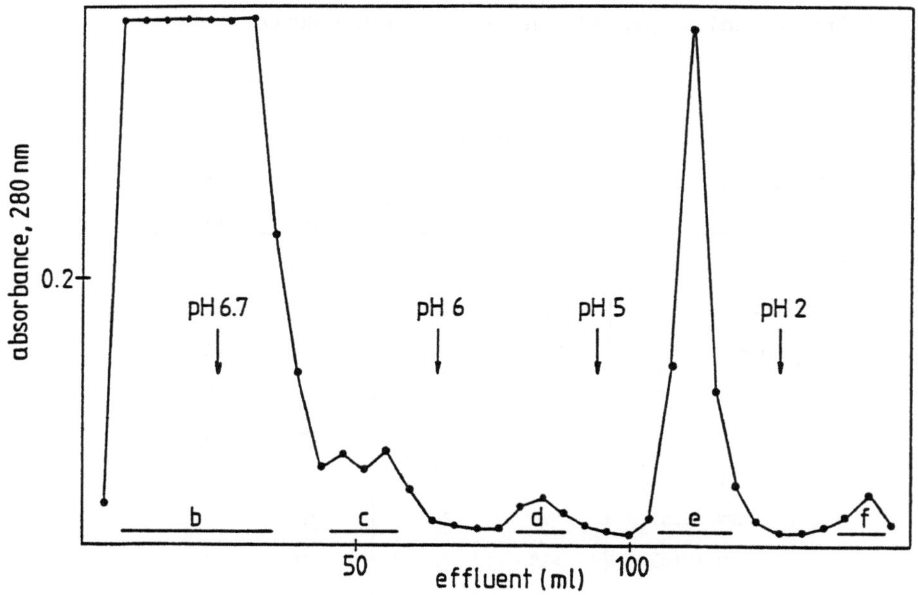

Fig. 12: Chromatography of $(His)_6$-DHFR
2 g E.coli cells were extracted with 6M GuHCl. After centrifugation, the supernatant was pumped onto a Ni^{2+}-NTA column equilibrated with extraction buffer. The column was developed with a pH-step gradient.
b: column flow-through
c: fraction eluted at pH 6.7
d: fraction eluted at pH 6
e: purified $(His)_6$-DHFR eluted at pH 5
f: fraction eluted with 0.2M acetic acid.

DHFR-$(His)_2$ extracted with 6M GuHCl was not retained on the Ni^{2+}-NTA column equilibrated with extraction buffer. Most of the fusion protein was found in the column flow-through.

In fig. 13 the chromatographic behaviour of the two described and eight additional poly-histidine fusions, examined under the same conditions, are summarized. These results show that the adsorption and elution behaviour of the poly-histidine fusions are strongly dependent on the solvent system used throughout the chromatography. For the purification in 0.05M phosphate buffer the di-histidine affinity peptide at the carboxy terminus was most efficient, whereas for the recovery in 6M GuHCl the hexa-histidine affinity peptide gave the best result.

Fusion protein	Phosphate Retained (%)	Phosphate Eluted (%)	GuHCl Retained (%)	GuHCl Eluted (%)
$(His)_2$-DHFR	30	10	< 5	-
$(His)_3$-DHFR	90	75	<10	-
$(His)_4$-DHFR	>90	30	10	10
$(His)_5$-DHFR	>90	20	50	50
$(His)_6$-DHFR	>90	10	>90	90
DHFR-$(His)_2$	>90	90	< 5	-
DHFR-$(His)_3$	>90	80	<10	-
DHFR-$(His)_4$	>90	50	10	10
DHFR-$(His)_5$	>90	40	50	50
DHFR-$(His)_6$	>90	30	>90	90
mDHFR*	< 5	-	-	-

Fig. 13: Affinity of poly-histidine fusion proteins for the Ni^{2+}-NTA adsorbent in 6M GuHCl and 0.05M phosphate buffer. The proteins were extracted and chromatographed as exemplified for DHFR-$(His)_2$ and $(His)_6$-DHFR. The amount of extracted poly-histidine fusion protein was set = 100%. Protein was estimated on coomassie blue stained SDS-Page.

Cleavage Of The -$(His)_2$ Affinity-Peptide With Carboxypeptidase A (CPA)

CPA releases most rapidly amino acids with an aromatic or large aliphatic side chain from the carboxytermini of polypeptide chains. When purified DHFR-$(His)_2$ was treated with CPA, two equivalents of His and one equivalent of Ser were released. This result is in good agreement with the expected release of amino acids according to the carboxyterminal amino acid sequence

••••••Lys-Gly-Ser-Arg-Ser-His-His

of the fusion protein, since Arg is expected to be digested very slowly with CPA.
After the enzymatic cleavage, the reaction mixture was pumped without further treatment on the NTA column used for the purification of DHFR-$(His)_2$. The protein was not retained on the affinity adsorbent, indicating that the -$(His)_2$ affinity-peptide was removed. The protein in the column flow-through was pooled and identified as DHFR by SDS-Page analysis. The specific activity of the pooled DHFR was 9.8 U/mg.

DISCUSSION

We have shown that fusion proteins comprising a protein of interest and a poly-histidine peptide can be purified very efficiently on a novel metal chelate adsorbent when charged with nickel ions (Ni^{2+}-NTA adsorbent). The efficiency of the poly-histidine peptide, as exemplified with DHFR, is dependent on the solvent system used throughout the chromatography. DHFR-$(His)_2$ was recovered with a purity of >90% and 88% yield in 0.05M phosphate buffer, but the same protein did not bind to the Ni^{2+}-NTA column in the presence of 6M GuHCl. On the other hand $(His)_6$-DHFR was purified very efficiently (purity >90%, yield:

90%) in 6M GuHCl, whereas this fusion protein bound too strongly in 0.05M phosphate buffer and could not be eluted completely. These results suggest that the di-histidine at the carboxy terminus for physiological buffers and the hexa-histidine - either at the carboxy or the amino terminus - for highly dissociating solutions are the affinity-peptides of choice.

Recombinant proteins expressed in E.coli often are formed in an unsoluble form. These refractor bodies can only be dissolved with detergents or highly concentrated dissociating agents as guanidine or urea. However, to most of the conventional chromatographic methods these solutions are detrimental. The method of choice may be chromatography in 6M GuHCl on the Ni^{2+}-NTA column in combination with the hexa-histidine fusion.
To split off the affinity peptide from the protein of interest, a specific chemical or enzymatic cleavage site may be introduced at the junction (1). However, this method has limitations. If the desired protein already comprises the recognized amino acid or peptide sequence, the product will be degraded. In the described example with DHFR-$(His)_2$ we exploited the presence of an Arg at the penultimate position to the carboxy terminus. Since Arg is not digested by carboxypeptidase A, the histidine affinity-peptide as well as the carboxy terminal Ser were removed. From the above discussion it is evident that the method of cleavage must be determined on a case-by-case basis, after careful analysis of the amino acid sequence of the protein of interest.
The observation that two additional histidine residues at the carboxy terminus have only a small effect on the enzymatic activity of DHFR (for DHFR a specific activity of 10.3 U/mg was measured compared with 9 U/mg for DHFR-$(His)_2$) suggests that the di-histidine fusion has minimal effect on its tertiary structure.
Therefore for many applications the removal of the poly-histidine

peptide might not be necessary. Poly-histidine fusions may well be suited for direct immunisation. On the other hand the observed 75% decrease of activity with $(His)_6$-DHFR indicate that the influence of the affinity-peptide on the biological activity of the protein cannot be predicted. In conclusion, we believe that the described method is applicable on large scale to a variety of proteins synthesized in bacteria, yeast or mammalian cells. Beside DHFR fusion proteins several surface antigens of Plasmodium falciparum have successfully been purified using this principle (14).

Recently a similar fusion concept for the purification of proteins was published (15). For the purification of a hybrid protein comprising of a His-Trp chelating peptide fused to proinsuline, immobilized iminodiacetic acid charged with Ni(II) ions (Ni^{2+}-IDA adsorbent) was used.

REFERENCES

1 Moks, T., Abrahmsen, L., Oesterlöf,B., Josephson, S., Oestling, M., Enfors, S.O., Persson, I., Nilsson, B. and Uhlen, M. (1987) Bio/Technology 5: 379-382

2 Germio, J. and Bastia, D. (1984) Proc. Natl. Acad. Sci. USA 81: 4692-4696

3 Sassenfeld, H.M. and Brewer, S.J. (1984) Bio/Technology 2: 76-81.

4 Porath, J., Carlsson, J., Olsson, I. and Belfrage, G. (1975) Nature (London), 258

5 Porath, J. and Olin, B. (1983) Biochemistry, 22: 1621

6 Lönnerdal, B. and Keen, C.L. (1983) J. Appl. Biochem., 4: 203

7 Sillen, L.G. and Martell, A.E. (1968) Special publication No. 25. The Chemical Society London

8 Hochuli, E., Döbeli, H. and Schacher, A. (1987) J. Chromatogr. 411: 177-184

9 Sulkowski, E., Trends Biotechnol. (1985) 3: 1

10 Li, S.S.-L., Fitch, W.M., Pan, Y.-C.E. and Sharief, F.S. (1983) J. Biol. Chem. 258: 7029

11 Bujard, H., Gentz, R., Lanzer, M., Stüber, D., Müller, M., Ibrahim, I., Haeuptle, M.-T., Dobberstein, B. (1987) Methods in Enzymology 155: 416-433

12 Chang, A.C.Y., Nunberg, J.H., Kaufman, R.J., Erlich, H.A., Schimke, R.T. and Cohen, S.N. (1978) Nature 275: 617-624

13 Hochuli, E., Bannwarth, W., Döbeli, H., Gentz, R. and Stüber, D. (1988) Bio/Technology, submitted for publication

14 Gentz, R., Certa, U., Takacs, B., Matile, H., Döbeli, H., Pink, R., Mackay, M., Bone, N. and Scaife, J.G. (1988) EMBO J. 7: 225-230

15 Smith, M.C., Furman, T.C., Ingolia, T.D., Pidgeon, C. (1988) J. Biol. Chem. 263: 7211.

STRUCTURE-ACTIVITY RELATIONSHIP IN BIOMEMBRANES INVESTIGATED BY INFRARED-ATR SPECTROSCOPY

U.P. Fringeli

Institute for Physical Chemistry, University of Vienna, Währingerstrasse 42, A-1090 Vienna and Swiss Federal Institute of Technology, ETH-Zentrum, CH-8092 Zurich

Abstract: Infrared attenuated total reflection (IR-ATR) spectroscopy is shown to be a sensitive tool to get structural information in aqueous environment on a molecular level. The technique is optimum if the sample is immobilized in an oriented way. In this paper we report on the formation of lipid bilayer membranes which are then used for interaction studies with local anesthetics and as a support for the oriented adsorption of membrane fragments with incorporated Na,K-ATPase (Na,K-pump).

INTRODUCTION

The activities of many membrane components such as lipids, proteins (enzymes, receptors, ion channels, ion carriers, etc.) as well as connective tissue materials (collagene, polysaccharides etc.) are most of all closely related to their molecular structures. Investigation of these relations is important for a better understanding of fundamental processes in biomembranes. Moreover, knowledge of structure-activity relationships strongly facilitates the complex procedure of drug design in pharmaceutical chemistry. In this field the approach to structure activity relations often comes from the part of the substrate. Knowledge of the active conformation of the substrate would substancially support the design of agonists and antagonists. Last but not least success in this field of biomembrane research consequently would help to reduce experiments with living animals.

U.P. Schlunegger (Ed.)
Biologically Active Molecules

In order to get such information the technical requirements to the experimental method are extremly high. Most detailed information is obtained so far from x-ray-analysis [1], which however, requires crystallization of the membrane component and eventually co-crystallization with the substrate. NMR techniques [2] enable structure determination in aqueous environment. The method, however, may be limited in some cases by its sensitivity. A third method which is able to complement x-ray, NMR- as well as all the other structure determining techniques, shall be presented in this paper. It is infrared attenuated total reflection (IR-ATR) spectroscopy. This technique is optimum for structural in situ investigations of immobilized thin layers, such as model membranes, membrane fragments, protein mono- and submonolayers etc. The achieved sensitivity is high enough in order to detect $\sim 10^{13}$ carbonyl groups per cm^2 with stationary techniques ([4-7], and this paper) and even $\sim 10^{12}$ amide groups per cm^2 if modulation technique [8] is applied.

MATERIALS AND METHODS

(i) IR-ATR Spectroscopy

Attenuated total reflection (ATR) technique is achieved by placing the sample material in close contact with the optically transparent internal reflection element of higher refractive index and working above the critical angle θ_c. From Fresnel's equations it follows that an electromagnetic field exists in the rarer medium beyond the reflecting interface, even under conditions of total reflection. This field exhibits the frequency of the incoming light, but the amplitude falls off exponentially with distance z from the surface.

$$E = E_o e^{-z/d_p}, \qquad d_p = \frac{\lambda_1}{2\pi(\sin^2\theta - n_{21}^2)^{1/2}}$$

d_p denotes the depth of penetration. n_{21} stands for the ratio n_2/n_1 of the refractive indices of the rarer medium (n_2) and of the internal reflection element (n_1). λ_1 denotes the wavelength λ_{vacuum}/n_1 in the latter. It follows that the penetration depth amounts to the order of magnitude of the wavelength λ_1 of the infrared radiation. This allows IR absorption measurements of thin layers, provided optical contact between rarer medium (sample) and

ATR plate may be established, e.g., by direct deposition of mono- or multilayers. For more details on ATR techniques the reader is referred to [3,4,9].

Detection of small changes of the IR spectrum of a monolayer sample in aqueous environment requires on the one hand the use of multiple internal reflections (MIR), typically 20-50, and on the other hand a high accuracy in difference spectroscopy. In this respect we have achieved best results with dispersive [4-6,8,10] and Fourier transform (FT) instruments by using a single trapezoid MIR-ATR crystal (Fig.1) for sample and reference, as well.

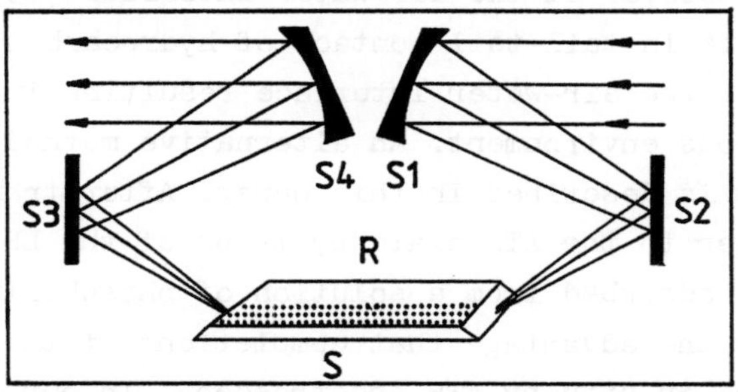

Fig.1. Single-Beam-Sample-Reference (SBSR) attachment. The parallel incident beam is focused on the entrance face of a MIR plate. The parallelism in the plate must be good enough to avoid crossing over between the sample (S) and reference (R) part. S1, S4: identical cylindrical mirrors. S2, S3: planar mirrors. The light is directed alternatively through the S and R side by means of a chopper. MIR: Multiple internal reflection.

Using parallel incident light which was focused to the entrance face of the crystal by means of cylindrical optics it was possible to place the sample on the lower (front) half, and the reference on the upper (rear) half. The IR beam was now passed alternatively through sample and reference side. In dispersive instruments a rotating chopper wheel with two rings with different numbers of holes was used in order to label sample and reference beams with different frequencies. In the case of FT instruments the beam was switched by means of a static chopper under computer control. Because sample and reference are measured with a single beam without cell displacement, this technique is referred to as Single Beam Sample Reference (SBSR) method. This special ATR attachment was installed in a BRUKER IFS 45 FT-IR spectrometer. Spectra were scanned with 2 cm^{-1} resolution. Spectra presented were smoothed under conditions which did not reduce the peak amplitudes of

CH-stretching bands.

(ii) Preparation Of Immobilized Bilayer Membranes

The preparation of a stable bilayer membrane on a solid substrate is not trivial, since the Langmuir-Blodgett (LB) technique [11,12] does not work reliably for bi- and multilayer transfer of phospholipids. After the transfer of the first monolayer, special techniques have to be applied for completion of the bilayer membrane, e.g. by moving the monolayer coated plate along the normal through a compressed monolayer at the air-water interface [13]. This procedure results in tail-tail contact of hydrocarbon chains on the plate and at the air-water interface resulting in a supported bilayer in aqueous environment. An alternative method leading to the same result is described in this paper. After transfer of the first lipid layer to the ATR plate by means of the LB technique, the second layer is adsorbed from a solution of phospholipid liposomes. This method has the advantage that completions of the second layer can be monitored in situ by ATR spectroscopy. In our application the first layer consisted of dipalmitoylphosphatidic acid transferred at 40 mN/m from the air-water interface of a LAUDA film balance (Lauda Königshofen, FRG) to both sides of the ATR plate. This plate was then mounted in an SBSR liquid sample cell. To transfer the second lipid layer, 1 mg/ml of 1-palmitoyl-2-oleoyl-sn-glycerophosphorylcholine (POPC) was suspended in phosphate buffer (20 mM) at pH 7.0. After 20 min of sonication at 35°C in N_2(g) atmosphere the solution was transparent. After cooling to 25°C it was filled into the sample (S) compartment, or for some applications in both, the sample (S) and reference (R) compartment of the SBSR cell. The POPC layer was completed after 30 min. Extension of the adsorbtion time to several hours resulted in POPC multilayer formation. Kinetic details will be published elsewhere.

(iii) Adsorption Of Membrane Fragments To Immobilized Bilayer Membranes

Membranes as described in sec.(ii) and [13] have the disadvantage that incorporation of substances into the membrane may be influenced by the solid substrate. For that purpose membrane fragments with enriched Na,K-ATPase (Na,K-pump) were adsorbed to an immobilized DPPA/POPC bilayer membrane according to the procedure described in [14]. Membrane fragments (~ 1 mg/ml) were prepared for adsorption in Tris/DCl buffer, pD 7.5. Details on adsorption kinetics as well as on the structural response of the Na,K-pump to Na^+ and K^+ environment, respectively, will be published elsewhere [15].

(iv) Chemicals

1,2-Dipalmitoyl-sn-glycero-3-phosphoric acid disodium salt (DPPA) and 1-palmitoyl-2-oleoyl-sn-glycero-3-phosphorylcholine (POPC) were obtained from R.Berchtold, Biochemical Laboratory, Bern, CH. Oxibutylprocaine (Novesin) was from SANDOZ-WANDER Pharma AG, Bern, CH. Membrane fragments with Na,K-ATPase prepared from rabbit kidneys were obtained from Dr.H.-J.Apell and Prof.P.Läuger, Fachbereich Biologie, University of Konstanz, FRG.

RESULTS AND DISCUSSION

(i) Immobilized Bilayer Membranes

The ATR plate was coated on the sample (S) and reference (R) side by a DPPA monolayer. This procedure resulted in a hydrophobic surface. Only the S side was contacted with a transparent solution of POPC liposomes until the DPPA/POPC bilayer was accomplished (cf. Materials and Methods). Then the liposome solution was replaced by an aqueous solution of 100 mM NaCl and 20 mM phosphate buffer pH 7.0. The same solution was introduced into the R compartment, contacting the hydrophobic surface of the DPPA monolayer. Fig.2a presents polarized IR-ATR transmission spectra of a POPC monolayer, the ratio between sample and reference (Fig.2b). Relevant absorption bands of lecithin are clearly resolved as can be verified by comparison with multilayer spectra [3] or aqueous dispersions [16]. Fig.2b shows the corresponding single channel spectra (S and R) with the very intense liquid water absorption bands (ν(OH):~3400 cm^{-1}, δ(OH): ~1640 cm^{-1}). The incompensation of δ(OH) in Fig.2a is predominantely due to the fact that the reference is coated by a monolayer whereas the sample consists of a bilayer. Accurate compensation of ν(OH) is no longer possible under these conditions

(Ge-plate, angle of incidence 45°, 30 internal reflections). Such bilayer membranes are stable over at least two days. They can be used for interaction studies, e.g. with drugs and as a support for adsorbed membrane fragments (see below).

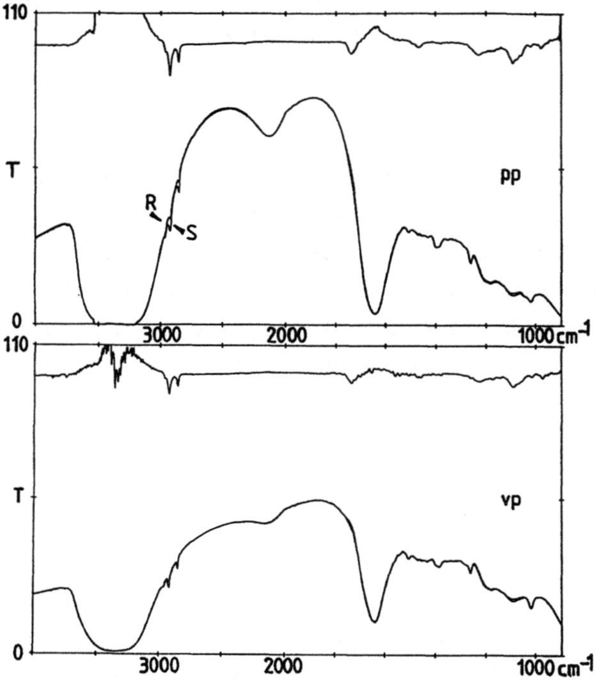

Fig.2. Polarized ATR spectra of a POPC monolayer attached tail/tail to a DPPA monolayer. The polar head groups of POPC are facing the aqueous buffer solution (100 mM NaCl, 20 mM phosphate). A DPPA monolayer in contact with buffer was used for compensation. Ge ATR plate, angle of incidence 45°, 30 active internal reflections. pp,vp: parallel and perpendicular incident light, T=20°C. R,S: reference, sample.

(ii) Interaction Of Localanesthetics With A POPC Membrane

Interaction studies of drugs [7, 17] or peptide hormones [18] with lipid model membranes have already been reported earlier. Adequate FT-IR equipment now enables such experiments to be performed in situ in H_2O environment, i.e. under conditions where in the common opinion IR spectroscopy was believed to have no chance to result in significant molecular informations.

Local anesthetics (LA) are suggested to react specifically from the cytoplasmic side with a receptor in the sodium channel [19,20]. Strong doubts with respect to this mechanism should be expressed, because of the chemical variety of LA, the required high clinical application doses (mM range) and the distinct tendency to associate and to adsorb to membranes in multilayers even at clinical application concentrations. Adsorption isotherms of 4 different LA have been reported in [7]. It was found that LA activity correlates with its tendency to adsorb to a POPC membrane.

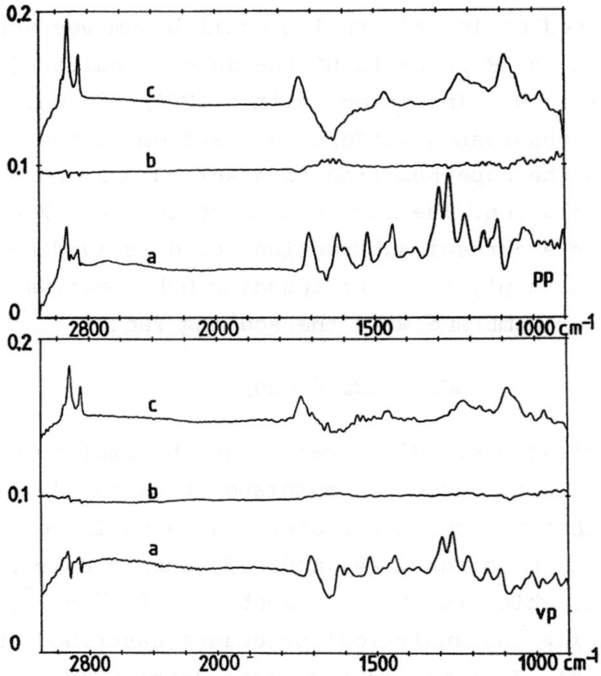

Fig.3. Interaction of a 0.3% (8.96 mM) solution (100 mM NaCl, 20 mM phosphate, pH 7.0) of the local anesthetic Oxibutylprocaine (OBP, Novesin) with a POPC model membrane.

a. Superposition of adsorbed (3-6 monolayers) and dissolved OBP in the bulk aqueous phase (about 1/3). Scale expansion: x3.

b. OBP solution in sample cell washed out by buffer solution 100 mM NaCl, 20 mM phosphat, pH 7.0). About 10 % POPC loss upon OBP detachment.

c. POPC absorbance monolayer spectrum (cf. Fig.2a) for comparison.

Fig.3 again demonstrates the quasi-reversible binding of the LA oxybutylprocaine (OBP, Novesin) to a POPC model membrane. It also demonstrates the quality of background compensation, since all spectra are real hardware difference spectra, i.e. the scaling factors of S and R spectra are unity and no base line correction was performed. A 0.3 % OBP solution (8.96 mM) was prepared in H_2O containing 100 mM NaCl and 20 mM phosphate buffer, pH 7.0. OBP concentration is typical for clinical application [7]. A DPPA/POPC bilayer membrane was prepared as described in sec.(i) above. Then the OBP solution was pumped through the S compartment of the cell in a closed cycle. Adsorption equilibrium was achieved within 10 minutes. In the R compartment the aqueous buffer (100 mM NaCl, 20 mM phosphate) was in contact with a DPPA/POPC membrane. The corresponding absorbance difference spectrum is shown in Fig.3a. It represents the superposition of adsorbed OBP and dissolved OBP in the bulk solution. The latter amounts to 30-40 % of the band intensity. For a correct subtraction, of dissolved OBP it is necessary to multiply the corresponding OBP spectrum obtained without bilayer membrane with the scaling factor

$$SCB = \exp(-2d/d_p) \qquad (1)$$

in both polarizations. SCB corrects for the smaller effective thickness due to the adsorbed membrane thickness d. It should be noted here, that the scaling factor indicated in ref.[7], eqn.(24) is not correct. Both eqns.(24) and (25) should be replaced by eqn.(1) above, which for bilayer membranes ($d \cong 50$ Å) is close to unity. Following the analytical procedure described in [7] it follows that Fig.3a reflects a surface density of $9.4 \cdot 10^{-10}$ moles OBP per cm^2. If the OBP solution in the S-compartment is replaced by the reference buffer solution, the desorption of OBP occurs within less than 1 minute. The equilibrium state of the model membrane after desorption, compared with its state before OBP adsorption is shown in Fig.3b. The lipid loss was 10 % as calculated from the $\nu_s(CH_2)$ difference band at 2850 cm^{-1}. Since one has to expect up to 90 % of surface coverage by OBP if the membrane is in contact with a 0.3 % OBP solution [7] one may conclude, that OBP adsorption is predominantely a surface effect because a larger lipid loss should be expected if OBP would penetrate significantly into the hydrophobic region of the membrane. More details on the orientation and the degree of protonation of adsorbed OBP will be published elsewhere.

(iii) Adsorbed Membrane Fragments Containing Na,K-ATPase

Starting again from a DPPA/POPC bilayer, membrane fragments were adsorbed from D_2O suspension to the POPC hydrophilic surface, as described in Materials and Methods above. The resulting system is expected to be represented by Fig.4. as was already suggested based on conduction measurements with black lipid membranes [21,22].

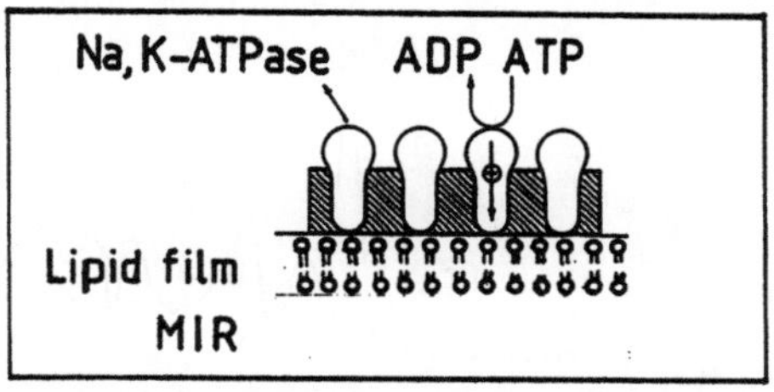

Fig.4. Expected structure of Na,K-ATPase incorporated into membrane fragments. The fragments are adsorbed to a DPPA/POPC bilayer. MIR: Multiple internal reflection plate.

Fig.5. represents one layer of membrane fragments in contact with Tris/DCl solution pD 7.5. From the intensity of the symmetric CH_2-stretching band $\nu_s(CH_2)$ at 2850 cm^{-1} [$\int OD_{\|} d\tilde{\nu} = 0.142\ cm^{-1}$, $\int OD_{\perp} d\tilde{\nu} = 0.094\ cm^{-1}$, $\int \varepsilon d\tilde{\nu} = 1.5\cdot 10^{6}$ cm/Mol, linear baseline] a surface concentration of $4.05\cdot 10^{-9}$ moles CH_2 groups per cm^2 is calculated [7]. Assuming 30 CH_2 groups per phospholipid molecule and neglecting the contribution of CH_2-groups from the aminoacid side chains the surface fraction of the lipid matrix of fragments (Fig.4) is found to be approximately 20 % (assumed lipid cross section 50 $Å^2$/molecule). Applying the same procedure to the Amid I,I' band at ~1640 cm^{-1} [$\int OD_{\|} d\tilde{\nu} = 1.832\ cm^{-1}$, $\int OD_{\perp} d\tilde{\nu} = 1.118\ cm^{-1}$, $\int \varepsilon d\tilde{\nu} = 6.3\cdot 10^{7}$ cm/Mol, linear baseline], one obtains $7.14\cdot 10^{14}$ amide groups/cm^2. Since the enzyme consists of two subunits with 1016 and 302 aminoacid residues [23] the Na,K-pump surface concentration results in $9.01\cdot 10^{-13}$ moles/cm^2.

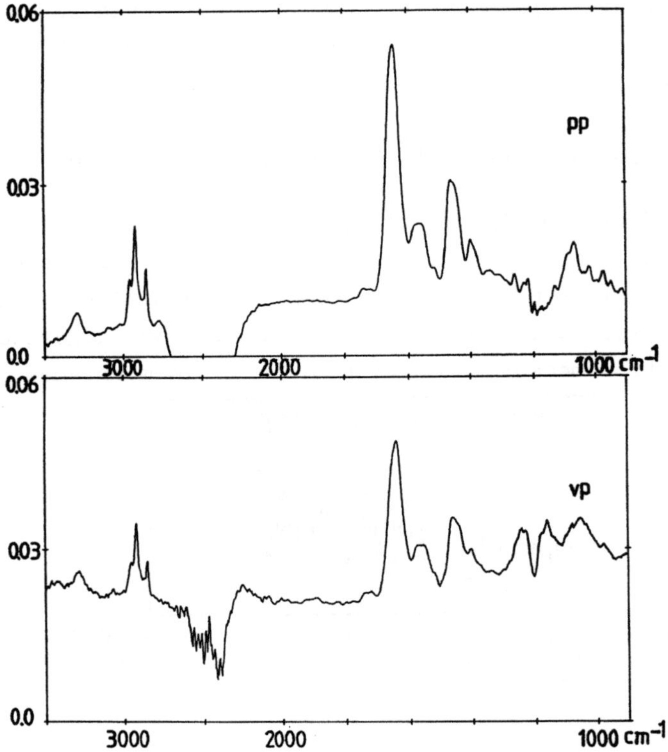

Fig.5. Polarized IR-ATR spectra of partially adsorbed membrane fragments with Na,K-ATPase in Tris/DCl buffer, pD 7.5. The area fractions of lipid and proteine are 0.2 and 0.12, respectively. Ge ATR plate,angle of incidence 45°, 30 internal reflections, T=20°C. pp,cp: parallel and perpendicular polarized incident light.

The band near 3300 cm^{-1} is typical for the NH-stretching vibration (ν(NH)) of the amide group. This band was resistent although H/D exchange was performed previous of adsorption and the adsorbed fragments were in $D_2O(\ell)$ contact up to 2 days. It follows that a certain amount of amide protons is not accessible for D_2O, a fact which was already observed from IR transmission measurements of a membrane fragment suspension [24]. Using an integrated molar absorption coefficient of ν(NH) of $4.87 \cdot 10^7$ cm/mole as determined from lysozyme [25], the amount of not accessible amide protons results in to be 29 %. These protons cannot be exchanged because of hydrophobic protection by the lipid matrix and/or because they are part of a stiff secondary structure. The polarization of the ν(NH) band $R = \int OD_{\parallel} d\tilde{\nu} \,/\, \int OD_{\perp} d\tilde{\nu} = 1.16$, favours the latter interpretation, because this value deviates significantly from the isotropic one, $R_{iso} = 1.67$ ($n_1 = 4.0$, $n_2 = 1.45$, $n_3 = 1.33$) [3,6]. From this dichroic ratio the mean direction of resistent NH-bonds is calculated to deviate at maximum 20° from the plane of the membrane surface. Such a polarization could originate from a helical segment oriented parallel to the surface, or from β pleated sheets spanning the membrane. The latter is expected to be more reasonable for the Na,K-pump. Similar membrane spanning β pleated sheet structures have also been reported with vinculin [6]. Finally, it should be mentioned that our findings qualitatively are in good agreement with the results of Brazhnikov et al.[24]. There are, however, some quantitative discrepancies which will be discussed in a forthcoming paper [15]. In this paper we also will report on the response of the Na,K-ATPase to K^+ and Na^+ environment, respectively.

CONCLUSIONS

IR-ATR spectroscopy was shown to be a sensitive tool for in situ studies with membranes and membrane components. In this paper methodological aspects of model membrane preparation have been discussed. We believe that especially the method of membrane fragment adsorption to supporting immobilized bilayers offers a wide field for biologically interesting experiments.

Acknowledgement

The author thanks to Ms.M.Fringeli for performing most of the experiments. Collaboration with Ms.J.Goette, Dr.M.Schöpflin and Prof.X.Perlia, Pharmazeutisches Institut, ETH-Zentrum,CH-8092 Zurich and with Dr.H.-J.Apell and Prof.P.Läuger, Fachbereich Biologie, Universität Konstanz, D-7750 Konstanz is very kindly acknowledged. I also thank the directorate of the laboratory for Physical Chemistry, ETH-Zentrum, CH-8092 Zurich, for facilitating experimental work. This research was supported by BRUKER, Analytische Messtechnik, D-7500 Karlsruhe, by the SANDOZ Foundation for Medical Research, CH-4002 Basel and by E.MERCK, D-6100 Darmstadt.

References

1 Deisenhofer J, Michel H, Huber R (1985) Trends Biochem Sci 10:243
2 Wüthrich K (1984) Biomedical Research Supplement 5:151
3 Fringeli UP, Günthard HsH (1981) In: Grell E (ed) Molecular Biology, Biochemistry and Biophysics; Membrane Spectroscopy. Springer, Berlin Heidelberg New York
4 Fringeli UP (1985) Nachr Chem Tech Lab 33/8:708
5 Fringeli UP (1980) J Membrane Biol 54:203
6 Fringeli UP, Leutert P, Thurnhofer H, Fringeli M, Burger MM (1986) Proc Natl Acad Sci USA 83:1315
7 Schöpflin M, Fringeli UP, Perlia X (1987) J Am Chem Soc 109:2375
8 Fringeli UP, Ahlström P, Vincenz C, Fringeli M (1985) SPIE Fourier and Computerized Infrared Spectroscopy 553:234
9 Harrick NJ (1967) Internal Reflection Spectroscopy, Wiley, New York
10 Fringeli UP, Günthard HsH (1976) Biochim Biophys Acta 450:101
11 Blodgett KB, Langmuir I (1937) Phys Rev 51:964
12 Kuhn H, Möbius D, Bücher H (1972) In: Weinberger A, Rossiter B (eds) Techniques of Chemistry. Wiley, New York
13 Tamm LK, McConnell HM (1985) Biophys J 47:105
14 Borlinghaus R, Apell HJ, Läuger P (1987) J Membrane Biol 97:161
15 Fringeli UP, Apell HJ, Fringeli M, Läuger P (1989) To be published
16 Casal HL, Mantsch HH (1984) Biochim Biophys Acta 779:381
17 Fringeli UP, Fringeli M (1979) Proc Natl Acad Sci USA 76:3852
18 Gremlich HU, Fringeli UP, Schwyzer R (1984) Biochemistry 23/9: 1808
19 Strichartz G (1985) In: Covino BG, Fozzard HA, Rehder F, Strichartz G (eds) Effects of anesthesia. Bethesda Maryland
20 Richie JM (1986) In: Roth SH, Miller KW (eds) Molecular and cellular mechanism of anesthetics. Plenum Medical Book Company, New York London
21 Apell HJ, Borlinghaus R, Läuger P (1987) J Membrane Biol 97:179
22 Fendler H, Grell E, Haubs M, Bamberg E (1985) EMBO J 4:3079
23 Shull GE, Lane LK, Lingrel JB (1986) Nature 321:429
24 Brazhnikov EV, Chetverin AB, Chirgadze YuN (1978) FEBS Let 93:125
25 Fringeli UP (1989) to be published

G. van Binst, University of Brussels, Belgium (Ed.)

Design and Synthesis of Organic Molecules Based on Molecular Recognition

Proceedings of the XVIIIth Solvay Conference on Chemistry
Brussels, November 28 – December 01, 1983

Chairmen of the conference: E. Katchalski-Katzir, V. Prelog

1986. 91 figures. XV, 315 pages.
ISBN 3-540-16123-6

Contents: Introduction. – Molecular Recognition in Biochemical Processes. – Thermodynamics, Kinetics and Stereochemistry in Molecular Recognition. – Synthetic Models of "HOSTS". – Design of new "GUESTS". – List of Authors.

One can learn a considerable amount from Nature, which uses molecular recognition to achieve selectivity in a degree so far unattainable to mere mortals. To analyze the structural features applied by Nature, to accomplish high molecular recognition, and to simulate these features by synthesis have recently become favorite occupations of chemists. This monograph contains the chairman's introduction, in which he summarizes the main points at issue, together with the contributions of the renowned scientists who participated.

Springer-Verlag Berlin
Heidelberg New York London
Paris Tokyo Hong Kong

Springer

MIX
Papier aus verantwortungsvollen Quellen
Paper from responsible sources
FSC® C105338

If you have any concerns about our products,
you can contact us on
ProductSafety@springernature.com

In case Publisher is established outside the EU,
the EU authorized representative is:
Springer Nature Customer Service Center GmbH
Europaplatz 3, 69115 Heidelberg, Germany

Printed by Libri Plureos GmbH
in Hamburg, Germany